Extractions

Global Ethics Series
Series Editor: **Christien van den Anker**, Reader, Department of Politics, University of the West of England, UK

Global Ethics as a field builds on longer traditions of ethical reflection about (global) society and discusses ethical approaches to global issues. These include but are not limited to issues highlighted by the process of globalization (in the widest sense) and increasing multiculturalism. They also engage with migration, the environment, poverty and inequality, peace and conflict, human rights, global citizenship, social movements and global governance. Despite fluid boundaries between fields, Global Ethics can be clearly marked out by its multidisciplinary approach, its interest in a strong link between theory, policy and practice and its inclusion of a range of work from strictly normative to more empirical.

Books in the series provide a specific normative approach, taxonomy or an ethical position on a specific issue in Global Ethics through empirical work. They explicitly engage with Global Ethics as a field and position themselves in regard to existing debates even when outlining more local approaches or issues. The *Global Ethics Series* has been designed to reach beyond a liberal cosmopolitan agenda and engage with contextualism as well as structural analyses of injustice in current global politics and its disciplining discourses.

Titles include:

Carlos R. Cordourier-Real
TRANSNATIONAL SOCIAL JUSTICE

Anna Grear
REDIRECTING HUMAN RIGHTS
Facing the Challenge of Corporate Legal Humanity

Shahram Khosravi
'ILLEGAL' TRAVELLER
An Auto-Ethnography of Borders

Ivan Manokha (*editor*)
THE POLITICAL ECONOMY OF HUMAN RIGHTS ENFORCEMENT

Darrel Moellendorf
GLOBAL INEQUALITY MATTERS

Michal Rachel Nahman
EXTRACTIONS
An Ethnography of Reproductive Tourism

Derrick M. Nault and Shawn L. England (*editors*)
GLOBALIZATION AND HUMAN RIGHTS IN THE DEVELOPING WORLD

Christien van den Anker and Ilse van Liempt (*editors*)
HUMAN RIGHTS AND MIGRATION
Trafficking for Forced Labour

Global Ethics Series
Series Standing Order ISBN 978–0–230–01958–4

You can receive future titles in this series as they are published by placing a standing order. Please contact your bookseller or, in case of difficulty, write to us at the address below with your name and address, the title of the series and the ISBN quoted above.

Customer Services Department, Macmillan Distribution Ltd, Houndmills, Basingstoke, Hampshire RG21 6XS, England

---

Extractions

An Ethnography of Reproductive Tourism

Michal Rachel Nahman
Senior Lecturer in Sociology and Criminology,
University of the West of England

palgrave
macmillan

First published 2013 by
PALGRAVE MACMILLAN

Palgrave Macmillan in the UK is an imprint of Macmillan Publishers Limited, registered in England, company number 785998, of Houndmills, Basingstoke, Hampshire RG21 6XS.

Palgrave Macmillan in the US is a division of St. Martin's Press LLC, 175 Fifth Avenue, New York, NY 10010.

Palgrave Macmillan is the global academic imprint of the above companies and has companies and representatives throughout the world.

Palgrave® and Macmillan® are registered trademarks in the United States, the United Kingdom, Europe and other countries.

ISBN 978–0–230–31929–5

This book is printed on paper suitable for recycling and made from fully managed and sustained forest sources. Logging, pulping and manufacturing processes are expected to conform to the environmental regulations of the country of origin.

A catalogue record for this book is available from the British Library.

A catalog record for this book is available from the Library of Congress.

Transferred to Digital Printing in 2013

For Ima, Abba and Lilach

Contents

Figures

Acknowledgements

How to thank all the people who supported, inspired and contributed towards my writing this book? There are many gaps and absences for which I apologize in advance.

The Wenner-Gren Foundation for Anthropological Research and the Social Science and Humanities Research Council of Canada made the project financially viable. I could not have completed the manuscript without the support of a University of West England Research Leave Fellowship and benefited greatly from my Visiting Senior Fellowship at that time (September 2010 to September 2011) at the BIOS Centre at the London School of Economics and Political Science. Participation in the BIOS Roundtable and a conference on reproduction and kinship there enriched my writing.

This book has also benefited from all of the feedback and reviews that I have received in the process of publishing journal articles and book chapters elsewhere. Parts of these are reprinted here with permission. These are: (2006) 'Materialising Israeliness: difference and mixture in transnational ova donation', *Science as Culture*, 15, 3: 199–213; (2007) 'Synecdochic ricochets: biosocialities in a Jerusalem IVF clinic', in S. Gibbon and C. Novas (eds), *Biosociality, Genetics and the Social Sciences: Making Biologies and Identities* (London: Routledge), pp. 117–35; (2008) 'Nodes of desire: transnational egg sellers as theorists of reproduction in a neo-liberal age', *European Journal of Women's Studies*, 15, 2: 65–82; (2010) '"Embryos are our baby": abridging hope, body and nation in transnational ova donation', in J. Edwards, P. Harvey and P. Wade (eds), *Technologized Images, Technologized Bodies: Anthropological Approaches to a New Politics of Vision* (Oxford and New York: Berghahn Books), pp. 185–210; (2011) 'Reverse traffic: intersecting inequalities in human egg donation', *Reproductive Biomedicine Online*, 23, 5: 626–33.

To the people in Israel and Romania who took the time – at one of the most intense times in personal and national terms – to talk to me with such depth and openness about egg donation and life in Israel and Romania, I am extremely thankful. Ofra Balaban was the best primary informant an anthropologist could want. I'm not sure that she would agree with everything that I have written here, but I gained many valuable insights from her. I learned much about ethics and reproduction in Israel from Carmel Shalev, Meira Weiss and Tsviya Birman. The doctors, nurses, cleaners, embryologists and administrators whom I spent time with in the clinics guided and opened their doors to me even when they didn't quite understand what a culture of science was, and I am indebted to them for this faith.

Sarah Franklin and Maureen McNeil supported me in this project from start to finish. Lucy Suchman, Iris Jean-Klein and Bob Simpson questioned and challenged me and their insights helped me carry it forward to make it into a book. Donna Haraway's incisive commentary on an early chapter draft helped shape my ideas around ova traffic. Celia Roberts, Michaela Fay, Fiona Summers, Rebecca Irvine, Tiago Moreira, Adrian Mackenzie and Adi Kuntsman read, listened and commented generously and warmly over the years. While at Lancaster, Jackie Stacey, Vicky Singleton, Claudia Castañeda, Gail Lewis, Anne-Marie Fortier, Nayanika Mookherji, John Law and Andrew Sayer were kind and willing readers and listeners.

Friends and colleagues more recently read and commented on chapters in their various incarnations, for which I am very grateful. Thanks go to Aditya Bharadwaj, Jeannette Edwards, Sarah Ellis, Maria Fannin, Zeynep Gurtin, Cathy Herbrand, Marcia Inhorn, Julie Kent, Audrey L'Esperance, Rosanna Machin, Mianna Meskus, Elizabeth Roberts, Jenny Reardon and Pete Wade. Past and present colleagues at the University of West England were extremely supportive, in particular Natalia Hanley, Lita Crociani-Windland, Alison Assiter, Sean Watson, Sean Creaven, Stella Maile, Havi Carel and Andrew Mathers. Andrew Mathers also took on much more than his fair share of the administrative duties of teaching during the final phase of writing, for which I am most grateful. PhD student Anna Petkovic supplied an extensive bibliography on exploitation and global ethics debates on organ trafficking. My editor Christien Van Den Anker believed in me and encouraged me to finish this project.

My Aunt and Uncle Hava and Yacov Nahman's ability to argue with me yet remain my surrogate parents in Israel despite our differences of opinion are evidence of their devotion and love. My Aunt and Uncle Perah and Avram Nahman always provide a warm, supportive, nourishing home in Holon. My cousins Liron, Tsachi and. Doron Nahman were amazing taxi drivers and friends. The memories of my Aunt Fanny Cohen and her daughter Ahuva Cohen-Ben Aharon pervade the book as women of strength and valour. The loving support I get from my sister Lila Beijer and my parents Maya Nahman and Israel Nahman is what has sustained me throughout the length of this project from start to finish. Dave Weltman and our Rosa made getting up in the morning to write an exciting and hopeful prospect.

Abbreviations

ART	Assisted Reproductive Technologies
AZH	Assisted Zona Hatching
CBRC	Cross-Border Reproductive Care
ERA	European Research Area
ICSI	Intra-Cytoplasmic Sperm Injection
IDF	Israel Defense Forces
IUI	Intra-Uterine Insemination
IVF	*In Vitro* Fertilization
MK	Member of Knesset (Member of the Israeli Parliament)
OHSS	Ovarian Hyper-Stimulation Syndrome
R&D	Research and Development
STS	Science and Technology Studies
TEFNA	Testicular Fine Needle Aspiration

Glossary

Aliyah	Ascent; also meaning 'ascent to Israel'
Ba-Aretz	In the land, meaning 'in the Land of Israel'; shorthand for referring to Israel in day-to-day talk
Dikkur	A pricking or stabbing. Used in Israeli clinics to refer to oocyte extraction
Kefiyya datit	Religious coercion
Kupa	Singular short form for healthcare provider
Kupat holim	Healthcare provider
Kupot	Plural short form for healthcare providers
Mash'eva	Pump
Mashgiha	A woman from the Orthodox Jewish community who is trained to oversee IVF treatment procedures to ensure that these comply with Jewish law. IVF patients sometimes privately employ them for this purpose
Mizug galuyot ideology	Assimilation; a blending of exiles or melting-pot
Piggu'a	Attack, usually a bombing or shooting
She'iva	Aspiration; suction
Takkala	Hitch (singular); *takkalot* (plural)
Te'udat zehut	Identity card; often referred to in singular, *te'uda*

1
Extractions

The 'terrorist' in the *in vitro* fertilization (IVF) clinic: notes from research in a private Israeli hospital

Behind Dr Shmuel's desk and to the right on the wall was a large map of the State of Israel (see Figure 1.1). It included but did not mark the Occupied Palestinian Territories and the Gaza Strip. It was a topographical map, with major roads and cities marked by dots and dashes. On this map, five round, coloured stickers were affixed, marking the location of the company's five satellite clinics in various parts of the country. At the top left-hand corner of the poster was a glossy company brochure. On this brochure was a picture of a blonde baby popping out of an oversized chicken's egg. The caption in Hebrew read: *'Lir'ot et ha-nolad, ve-maher!'* 'To see the [child that is] born, and fast!' On the wall next to this map was a photograph of four European-looking people: the doctor, a woman, a man and a baby, presumably two of the doctor's happy new parents with their baby.

Next to the photograph was a large whiteboard. I didn't take a photo of it for reasons that I think will become obvious. On it was a large drawing, in black marker ink, of two men. One, scruffy faced, with a large nose, overgrown beard and menacing eyes, wielding a large knife. In the context of the Al Aqsa Intifada that was then raging, this man could be none other than the Palestinian[1] 'Terrorist'. The other, clean-cut European-looking, was dressed in army fatigues with the acronym *Tsahal*, 'IDF' (Israel Defense Forces), written on the front of his shirt. He was holding a large rifle and had a speech bubble above his head. The writing in the bubble said: *''Am Israel hai,'* 'The nation of Israel lives.' This is a common religious phrase of exaltation and survival, which little children and other celebrants sing in Israel as in the

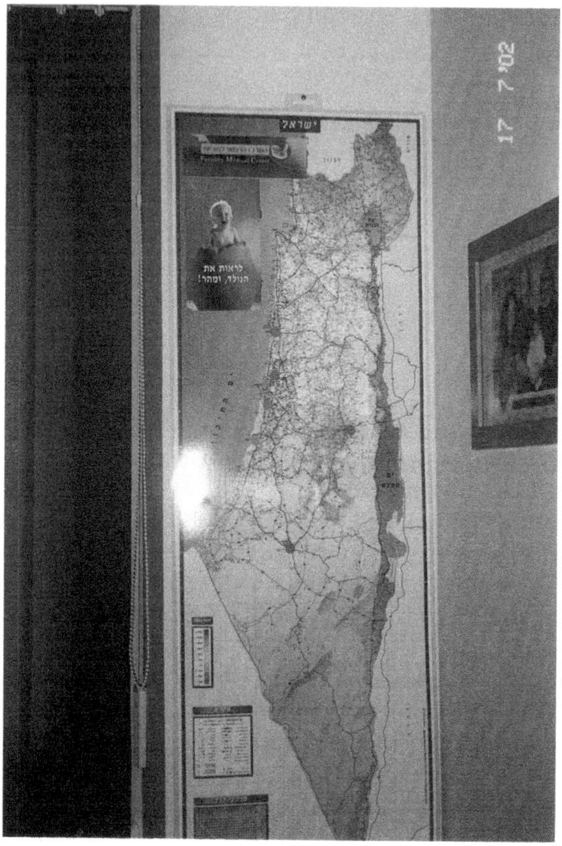

Figure 1.1 Map of Israel with baby and family
Source: The author.

Diaspora when marking Holocaust Remembrance Day. Here, however, it resonated as well with the prevailing sentiment that Jewish Israelis 'are fighting for survival' – what politicians, the media, egg recipients, my Israeli family and friends alike repeatedly called '*milhama 'al ha-bayit*' or 'a war over home'.[2] In the clinic, the statement written on the board, ''*Am Israel hai*', is one in which the survival of the European Jewish self, nation and state are rolled into one. During the brief period since the founding of the state in 1948 and including the time covered in this research, everyday violence was intimately co-inscribed by hetero-normative family values[3] and imaginaries of racial superiority of whiteness.

The very presence of images of the family and the nation in the clinic in Israel may be unsurprising, given how historically the reproduction of nations and states has been tied to the reproduction of the individual (Yuval-Davis, 1989, 1997) and given the banality of militarization in Israeli society. But here we have the 'Terrorist', in a private Israeli clinic where *in vitro* fertilization (IVF) is very costly.

There is an uncanny kinship between these images of the Israeli soldier and Palestinian terrorist confronting each other, and the blonde baby coming out of a chicken's egg, on a map of the State of Israel, next to a picture of a happy Jewish family. Something familiar and echoes of narratives of improvement, security, protection of nation/family and ultimately of a blonde child reside here. It is like the recent history of Ashkenazi Jews (Jews who self-identify or are identified by others as coming from Europe), conflated with Zionist state ideologies of victimhood and violence is written on the wall. This is about fears and anxieties over racial extraction, the removal of Palestinians from the land, and taking eggs out of women's bodies for profit. The images described above show just how nation, race, war, the individual self and egg donation were at once overlapping and mutually constituting 'domains'. As I argue throughout this book, the 'individual' stands in for the 'collective' in Israel. So too egg donation and the nation can stand in for one another. After observing these images I could not follow the conversation that the women in the room were having. I tried unsuccessfully to make sense of what I had just seen. When I asked around, I found out that the drawing was made by Dr Shmuel's ten-year-old son and his friend who were bored while spending a day at the clinic a few months earlier.

Extractions ethnographically maps and traces relations of removal, fertilization, transport and transfer of human eggs between women's bodies and clinics in Israel and Romania in 2002 (since 2006, Romania no longer permits the selling of eggs) at the height of the Al Aqsa Intifada (the Second Intifada). One year after 9/11, this was also a time of heightened concern with nationalism, security and borders in the global North. Oocyte extraction practices are indexical of this time. *Extractions* deals centrally with the questions: How do everyday practices and procedures of Israeli egg donation create and reinforce nationalized, racialized bodies, borders and ideals? How do we make the study of something as unfixed and complex as egg donation, race, war, gender and class coherent without shutting it down? How do we employ a critical sensibility for globalization and gender in this process? This book is an experiment in doing these things all at once.

Whilst being attentive to the influence of Jewish law on Israeli egg donation, the focus here is on other influences on the political and material dimensions of reproductive technologies. It is a concern with how practices and technologies of reproduction produce Israeliness and how Israeliness produces reproduction. Whilst this has been done before to some extent (Portugese, 1998; Kanaaneh, 2002; Weiss, 2002), the integral nature of the occupation and violence is brought from the margins of the narrative to its centre. 'Israeli extractions' (that is, identities and ova) get made in complex and messy ways that 'interfere' with both state and scholarly discourses about Israeli reproduction. Writing about this is also an 'interference' (Haraway, 2004). Gendered citizens are simultaneously situated along axes of religion, reproduction, class and race, in relation to official discourses about multiculturalism that are in themselves both celebratory and cautious. With overlaps, disjunctures and unlikely juxtapositions, the reader is asked to reflect critically on the ways in which we represent these things.

Transnational reproduction and migration are linked concerns, and there is a historical and ideological overlap in Israel between citizenship law and egg donation law. In the year 2000, it was clear that (despite disagreements) the authorities had decided that the donor's genetics did not matter for ova donation and for the importance of maintaining a Jewish family (Kahn, 2000). Yet over the ten years since then, it seems that a different perspective has been adopted in which the genetic/religious/racial make-up of the donated egg has gained importance. This has occurred both in response to the perceived threats to the state from non-Jewish Asian and African migrants and Palestinians, and the pressures for the Israeli state to be a global player in the marketization of cross-border reproduction.

Israeli borders are being hardened through the creation of laws concerning loyalty to the Jewish state. The right-wing Israel Beiteynu Party (translated as 'Israel Our Home') recently proposed a series of five 'citizenship equals loyalty' laws, which have been and continue to undergo readings in the Israeli Knesset (Parliament). One law proposes stripping Israeli citizenship from those who are unwilling to declare loyalty to the Israeli state. Whilst that particular proposed law was not approved, another version of the law has been approved by the Cabinet and is awaiting final approval by the Knesset. This law requires non-Jewish people wishing to become citizens of Israel to declare their loyalty to 'the State of Israel as a Jewish and democratic state'. In the year 2010, the Israeli government passed an Egg Donation Law that bans cross-religious donation, a move that seems to conflict with both the Israeli state policy of supporting a multicultural state and Jewish religious authorities' teachings on the kinship and citizenship significance of donor eggs.

There has been a genetic reinscription of race (El Haj, 2007), and the ethnographic research that I conducted in 2002 already bore witness to this (also this law, although only passed in 2010, was already written almost entirely by 2002). So, even if you are not born to a Jewish mother, you can be a citizen by reaffirming the centrality of Jewishness to the state. And in egg donation, you must have come from the egg of a Jewish woman to be counted as a legitimate Jewish citizen. Equally, you cannot have come from the egg of a Jewish woman if you were born to a non-Jewish woman, according to this new law. Affiliation to the state is being negotiated in hearts and genes. The state's lawmakers use Jewish kinship thinking in ways that shift according to the complex context. These are financial, reproductive and border/citizenship concerns all folded into one another. They often hinge in public discourse on ideas of the state needing to protect itself from the violence of Israel's enemies both present and past. Whilst the material gathered in this book predates these policy changes, they give evidence to the ways in which common sense and everyday forms of racism and border disputes fed into the contemporary global bioeconomy.

A central trope in lay and scholarly discussions about both citizenship and reproduction is 'pronatalism'. Ideas about the Israeli state's pronatalism and 'world leader' status may be a discourse that perpetuates an ideal of Israel as a Jewish, 'chosen' nation rather than a nation-state founded on expropriation of land, expulsion. These competing myths of the nation can also produce different narratives about reproduction. The peremptory assertion of Israel's pronatalism demands detailed microscopic unfoldings of how that process happens. *Extractions* attempts to unravel those enactments (Mol, 2002) of the body and the nation by switching between the microscopic stuff of individual bodies to the scale of the nation in a kind of 'synecdochic ricochet' (explained further below; also see Nahman, 2008; Hayden, 1995). Where the 'part' stands in for the 'whole' but like the bullets and bombs that Israelis and Palestinians were dodging at the time of this research, there is a constant move back and forth between these so-called 'domains' of knowledge and practice. This is a historicized feminist practice of writing. Exchanges in eggs, talk about eggs, legislation regarding the importing and donating of eggs, and processes of extracting, fertilizing and freezing eggs that occur in disparate places, and in specific moments in time perform (undo and reify) Israeli nationalism, identity and borders. In nationalism and nationalistic practices, nations turn particular symbols into synecdoche. The hijab standing in for the Muslim woman and the so-called dangers of Islam to Europe is one example (Bowen, 2007, cited in Vertovec, 2011). If nation-states use synecdoche as a way of imagining 'the other' and therefore of constituting themselves, then using synecdoche to analyze nationalism is using an emic (or indigenous) term in the study of nationalism. I will return to this below.

In the social sciences, 'the body' has been studied as a metaphor for a society's concerns about its borders. In earlier anthropological accounts, these concerns shaped cultural practices around purity, dirt and contamination (Douglas, 1966). Structuralist, symbolic and psychoanalytic arguments were used to explore these relationships. In contrast, social classes (and nations as larger examples of these) were seen to be produced through the repetition of daily bodily praxis (Bourdieu, 1977). For Douglas, the relation between the body and the community is created through the symbolism between structures and for Bourdieu it is created through the repetition of everyday practices that generate social structures. Both perspectives allow for the idea that aspects of the body transmit meaning about borders of the collectivity and that, in turn, the body's meaning is shaped by the larger collectivity. Nevertheless, the micro processes of the body through which this larger collectivity is formed have been seen in the work of feminist accounts of the body (Martin, 1991; Ginsburg and Rapp, 1991, 1995; Franklin, 1997). *Extractions* contextualizes these practices within the global markets that are opening up to facilitate reproduction.

Race, IVF and parenthood have been examined in the USA and UK (Thompson, 2005; Wade, 2007), while anthropology has dealt extensively with broader concepts of reproduction and notions of exchange in general (Konrad, 2005; Strathern, 1988, 1995). Studies of IVF and reproduction in India (Bharadwaj, 2006a, 2006b; Gupta, 2006) have questioned the connections between neoliberalism, the state, religion and reproduction. Scholars studying the Middle East (Tremayne, 2006; Inhorn, 2003; Zuhur, 1992) have analyzed relationships between Islam and reproduction. Embodiment and medicalization have been central concerns of recent ethnographers of Israeli reproduction (Teman, 2010; Ivry, 2009). Medical 'tourism', cross-border reproductive care and medical migration have been gaining significant scholarly attention from within anthropology, law, sociology, bioethics and other fields (Roberts and Scheper-Hughes, 2011; Inhorn, 2011; Gürtin, 2011; Shenfield, 2011; Storrow, 2011; Dickenson, 2007).

Israeli IVF has been examined, with reference to the everyday practices and the rabbinical and state discourses around 'Jewish kinship' (Kahn, 2000). Along with other Israeli scholars I argue that there are many ways in which reproductive technologies in Israel are tied to 'the nation', particularly through militaristic metaphors (Ivry, 1999, 2009), and through cultural ideas of nature and motherhood (Teman, 2003, 2010). *Extractions* ethnographically extends this field and looks at the political, economic and ethical dimensions of the unequal positions of women in differently positioned societies, and illustrates how reproductive technology relates to the contemporary context of Israeli state politics. The discursive practices of ova donation that I observed create novel relationships between the individual, reproduction and the Israeli state. With its attention to global economic and Israeli political 'cultures', this book thereby contests the rather more exclusive focus on Jewish ideals of pronatalism.

This book presents Israeli accounts of eggs as a resource that is in 'shortage'; one that can 'make more disposable people'; and is narrated as being forcibly removed (through *dikkur,* stabbing or poking); 'returned' into different women's bodies; embodying a 'promise' for a child; something explosive; and providing sources of income for donors and clinicians. The research phase of this project involved a nine-month study conducted in Israel and Romania between January and September 2002. Qualitative methods of participant-observation in Israeli IVF clinics, and open-ended interviews with doctors, nurses, Israeli (Jewish and Palestinian) ova recipients, (Israeli and Romanian) ova donors, lawmakers, lawyers and patient advocates were undertaken in the data-gathering phase of the research. Open-ended, semi-structured interviews were carried out with 25 ova recipients and recipient couples, 2 Israeli ova donors, 20 Romanian ova donors, 5 Israeli IVF doctors, 3 representatives of the Israeli Ministry of Health (one of whom is a rabbi and gynaecologist and two of whom are lawyers), 2 egg donation activists, a lawyer who won a key case at the High Court of Justice for importing Romanian eggs, and numerous other healthcare professionals and clinic staff. I analyzed this material qualitatively, looking for emerging patterns in how ideas about egg donation were being discursively produced. The arguments elaborated below emerge directly out of this research. *Extractions* examines both Jewish and Palestinian experiences of assisted reproduction in Israel. It does so at the juncture of the transnational movement of reproductive substances.

Anthropology makes culture: on the particular and the universal

Israeli egg donation is costly, people must be able to pay initially out of pocket for treatments, are urged to undergo as many possible treatments as necessary and subsequently often fall into debt. These are classed and racialized outcomes of the strong push to reproduce in Israel. A racial topography of egg donation is found in Israel that is similar to the racialized practices of egg donation in other Western contexts. This similarity is crucial because the anthropological endeavour to find 'the particular' can sometimes make us the handmaidens of hegemonic nationalisms. A preference for analytic attention to the particulars of Jewish history and the inception of the Israeli state can lead to an over-attribution of these *particular* aspects to Israeli reproduction as a whole. The focus on histories of anti-Semitism, religious division between Jews and non-Jews, or even the secular–religious split amongst Jews to the exclusion of other aspects (internal racism, internal class divisions, cross-national similarities with other Western countries) can inadvertently *reproduce* the tropes and myths of the founding of the state rather than merely explain them.

For instance, Jewish Israeli women accept and even desire ova from Romanian women, regardless of the religion of the donor and recipient because they are white and European (Nahman, 2006). Conversely, Jewish Israeli women will rarely admit that they would accept ova from Palestinian women. In addition, the everyday practices of Israeli IVF clinics have been shown to reinforce heteronormative values, even as the state attempts to 'pinkwash' the occupation by promoting its pro-gay stance. Transnational Israeli practices of oocyte extraction, transport and transfer (the technical term for inserting an embryo into a woman's uterus) practices embody the Israeli concern over borders and are, in themselves, a way of allowing Israelis to 'stay put' given the transnational mobility of eggs. One might expect that transnational ova donation practices reproduce the State of Israel as 'globalized' and 'transnational'. And they do. But they also reproduce ideas and practices of the Jewish nation-state *through* practices of transnational ova donation. These practices involve movement across national borders that reproduce Israeliness as Europeanized and Jewish, reinforce the banality of crisis and allow subaltern Israelis to identify and resist the national body in its hegemonic form. In all of this, Romanian women are not just an exploited global resource, but also active participants who inter-cumulate. The practices of egg donation described here in their minutiae are telling a story about national aspirations, politics of ethnic cleansing, eugenics and everyday crisis, violence and victimhood.

A layer of Israeli inequalities is that which despite official laws in Israel that deem all of its citizens to be equal, the realities of everyday life dictate that one must be Jewish in order to be a legitimate citizen, with equal access to good jobs, healthcare and education (Dominguez, 1989; Peled, 2004; Liebman and Don-Yehiya, 1983). It has been suggested that this social pressure emerges out of the anxiety regarding the perceived demographic threat to the existence of the Jewish state posed by the reproduction of Palestinians (Portugese, 1998; Yuval-Davis, 1989, 1997). It has also been argued that the pressure to reproduce is so strong that Jewish religious law and consequently Israeli state law are among the most permissive in the world for access to reproductive technologies (Kahn, 2000). This has been identified as being part of the practices of creating a 'chosen body' (Weiss, 2002). The idea of a chosen body derives from the biblical notion of the Jews as the 'chosen sons' of God. It is a central trope of Zionist ideologies (Almog, 2000). According to Israeli anthropologist Meira Weiss, the 'chosen body' in Israel is an imagined ideal that is Jewish, Ashkenazi (European descended), militarized, masculinized and reproductive. Yet, whilst the ideal national body is Ashkenazi, a majority of Israelis today self-identify with Jewish-Arab and African diasporas, as well as with Muslim and Christian Palestinian national identities (with many additional complex and intersecting dimensions). *Extractions* suggests that the chosen body is a shape-shifter, a cyborg, imagined as male but selectively including and then excluding straight women, lesbians, gay men, bisexual people, trans- people and the myriad ways in which humans express themselves physically and sexually. It also excludes women who do not wish to or cannot reproduce. Several of my research participants at a fertility clinic said that they were 'second class' citizens due to their lack of a child. These are some of the contradictions of Israeliness. The promise of embryos and eggs in this context then is their enabling power to make 'chosen bodies' out of some 'unchosen' ones.

The Israeli state's relationship to the outside world doesn't only relate to issues of anti-Semitism or being a state under attack. Its relationship also has to do with its very powerful participation in global markets. As examples in this book about transnational buying and selling of human eggs show, the Israeli nation is made transnationally. Global markets in which Israel is a successful player, due in large part to heavy American subsidy, which is invested in research and development in science, technology and the military, are also central to this oocyte extraction story. Seeing how these practices are embedded within global circuits of exchange is important. The Israeli state operates very much as part of a global economy and culture (Nitzan and Bichler, 2002).

The account provided here illustrates how a particularly narrowly conceived Israeliness is being reproduced by the transnational 'flow' of eggs. Whilst it might be tempting to suggest that because of the transnationality of trade in gametes 'Israeliness' is more porous and cosmopolitan, such a conclusion would involve a misplaced 'biologization' and much historical forgetting. Acquiring ova from non-Jewish women in Romania does not alter notions of Israeli national identities just because the eggs come from non-Jewish women. That is, my research found Israelis quite willing to receive biogenetic substance from non-Jews and still to confer Jewish identities on their resultant children. Indeed, 'mixture' and 'difference' are central to Israeli nationalism. As one leading member of the Israeli Ministry of Health stated when I interviewed her:

> [T]here will no longer be Sephardim [Jews who follow Eastern or Judeo-Spanish religious traditions], Ashkenazim [...] everything will become mixed [...] the truth is today it's a lot less problematic [ethnicity in Israel] [...]. There's no social problem between the *edot* [ethnic groups]. (Ministry of Health, Israel, interview, 24 June 2002)

Yet, it is clear that despite this governmental perspective, the State of Israel is a cosmopolitan, multicultural society that experiences significant racial inequality among Jews. In these exchanges of ova, the ideals of Israeliness do not shift towards an enduring or transformative desirability of Africanness, Arabness, Asianness or Russianness, to note just some diasporic and marginal kinds of Israeliness (Alcalay, 1993; Lavie, 1996; Peled, 2004; Shohat, 1989). Rather, the transnational flow of eggs involves techniques and practices, which reinscribe the idea of the imagined 'need' for borders.

Whilst stating that Israel is pronatalist may work to critique biopower (state management of populations) in its particular version in Israel, it does not necessarily mean that one is criticizing ethnic nationalism or racism of the state. I too have represented Israeli Assisted Reproductive Technologies (ART) as being in line with the Jewish edict to 'be fruitful and multiply' (Nahman, 2005). Some authors indicate that this 'multiplying' is seen to be necessary due to the State of Israel's geographical position amid hostile surroundings. Yet the critique does not often go so far as to look at the Israeli state's own hostile economic, military and political activities. The critique more often stays in line with a liberal attitude that the state should not manage the bodies of populations – or as asked in one of the Israeli mainstream newspaper articles on this subject, 'What is the state doing in the bedroom anyway?' This perspective of how biopower (Foucault, 1978) is enacted on the bodies of national subjects is a worthy and important critique. Yet it does not go so far as to situate the Israeli case of ART within a wider historical and political context (with a few notable exceptions, discussed earlier).

These particulars *are* universal. Whilst they require an attention to local histories, those are not divorced from European histories. They are not 'inherently' Jewish or Israeli but rather about how global capitalism and European history transpire in this particular time and place. Attention to the actual historical events and crises that occurred during fieldwork enables an account that critiques national myth-making, whilst making its own counter-story. Important as it is, this 'local' dimension must not make us overlook the other equally salient aspects of Israeli histories, identities and bodies that may be equally interesting and relevant to an anthropological analysis of IVF and egg donation in Israel.

Sensing four kinds of extractions

The title *Extractions* highlights the four central aims of my account. First, is the descriptive analysis of the practices involved in procuring human eggs. Second, the extraction of profit from women's bodies viewed from the perspectives of the donors, the recipients and clinicians at a private IVF clinic. Third, the politics of race are seen as a central structuring force in the Israeli state and national imaginary through choices of egg/donor/baby traits and through everyday practices that racialize people. This undermines the idea that Israeli reproduction is solely about Jewish versus non-Jewish or religious versus secular. Fourth, *Extractions* represents a cyborg feminist writing method of extracting and overlapping parallel narratives about bodies, borders, race and war, to see what these narratives say about one another and about the world in general.

Oocyte extraction

I take as my starting premise the fact that oocyte extractions, just like gender and race, are not self-evident things in themselves. They have a politics that involve relations of power and of production. Oocyte extractions are discursive practices (Butler, 1993), sets of actions and ways of speaking about the world that bring them into being. They are material-semiotic generative nodes (Haraway, 1997), matter and meaning at once, which are enacted or brought into being (Mol, 2002). To separate them out from all the aspects that materialize them would be to deny how things are made through multiple entanglements (Barad, 2007). I realize that each of these ways of naming what I am trying to say has a slightly different intellectual genealogy, bringing them together is an act of making a monstrous child of anthropological, feminist philosophical and Science and Technology Studies (STS) ways of knowing about the world.

Taking eggs out of women's bodies is a process that looks and sounds different in different places. In many Western countries this removal is called an 'aspiration', a kind of gentle suctioning, like breathing, light and easy and hopeful. In Israeli clinics the word '*sheiva*' or suctioning was often used when talking to patients. More often, however, in the actual surgical theatre I came across the word '*dikkur*' meaning a stabbing or pricking. This language structures the practices to a certain degree. The process itself involves inserting a long ultrasound probe and needle into a woman's body vaginally and gently poking at the ovaries until the membrane is punctured and follicular fluid is sucked into the syringe. In Romania where I observed oocytes being extracted from the bodies of paid egg sellers, this was occasionally a bloody process where things went wrong, both due to the lack of specialized training of the doctor performing the procedure and because of machines breaking down. Women were monitored by a trained anaesthetist and were under a light general anaesthetic. These women and their clinicians (all young women as well) were doing something different – changing history in a sense for Romanian women, rather than being tools in a national project as women often were under the Ceausescu regime, they were turning themselves into resources for cumulating, making money, getting advancement in a newly Westernizing state.

In Israel, women undergoing IVF having oocytes extracted were generally being treated by trained, specialized IVF physicians, monitored by highly efficient machinery monitoring their oxygen levels. It was much less harrowing to watch, but the everyday world all around us was full of crisis and explosion. At the height of the research in March 2002, Palestinian civilians were regularly being killed by the Israeli army, and Israeli civilians were being killed by lone Palestinians who'd crossed what is euphemistically called a border between Israel and the Occupied Palestinian Territories. Oocyte extraction is about these kinds of entanglements.

The business of transnational extractions

The State of Israel is often represented as having the highest number of fertility clinics per capita in the world, being extremely pronatalist and a 'world leader' in the provision and practice of reproductive technologies.[4] Indeed, there is a high level of state support for IVF, although less so for egg donation. The state subsidizes egg donation (within Israel, not transnationally) for women aged 45 to 51. This subsidy is unlimited until the conception and birth of two children (Rabinerson et al., 2002). A study conducted of this age group across two years with a sample of 254 Israeli women found that the 'success rate' (meaning the number of babies born) was 17.7 per cent (ibid.). Despite these seemingly low success rates, Israel continues to subsidize ova donation for all its citizens. Egg donation more specifically has been the subject of intense longstanding debate both in the Knesset and outside it among feminists, legislators, healthcare providers and insurers and religious figures. A central outcome of these debates has been an Egg Donation Law passed in 2010, 'liberalizing' Israeli egg donation. Now any woman in Israel may 'donate' her eggs for approximately NIS10,000 (about US$2127).

The State of Israel is a major player in the global arena of science and technology research and development (R&D). Indeed, it has been recognized as a research 'tiger' that has withstood financial collapses around the world, owing in part to its wealth of research and development workers (Arora and Gambardella, 2005). It is especially indebted to USA public and private investment in its military and other technologies, indeed the State of Israel receives a third of USA aid globally, which it is heavily dependent on. It is a participant in the European Economic Community and as such a leader in European biotech research and development. It has been argued that its economy has always been transnational as its economy survives mainly from heavy subsidies by American private and public institutions as well as European multinationals. Healthcare is a major pillar of this neoliberal (market-oriented) global economy and the State of Israel is a particularly useful example for reflecting on the global phenomenon of the introduction of market concerns into health and healing (Filc, 2005, 2009). Indeed, Israeli healthcare, which was once state-funded, shifted in the 1990s to being partly privately run, resulting in a two-tiered health system.

There is a vibrant feminist debate about the bioethical dimensions of oocyte donation from women in poor countries. Often this work has made some very important points that women are exploited and doctors are making a profit. However, in this debate very little attention has been paid to what the women and doctors actually say, how they construct these notions of the extractability and saleability of the body. This matters, not as a way of excusing it or of saying that women make money and that they should be making more money or doctors are making money and they should be making less money. This isn't about a 'value in' 'value out' debate that bioethicists such as Donna Dickenson (2007) and Heather Widdows (2009) have been having. Rather, it is about making sense of the specificities of the experiences and narratives of extraction, but inserting the voices of egg sellers, both women and doctors is an important contribution. The book begins with a chapter on the Romanian egg sellers as theorists of reproduction. These women are generally young and willing to undergo several rounds of extraction. They are given hormones over a month-long period in order to stimulate their ovaries to produce around twenty to thirty eggs, an enormous number for a normal woman who would usually produce only one a month. They are paid about US$200 for their eggs. Compared to other forms of labour in Romania, this is generally seen as a relatively easy thing to do. Wages in Romania are extremely low compared to the rest of Europe and labour relations are such that the majority of workers do not enjoy a great amount of autonomy.

Israeli doctors making transnational reproduction available to Israelis through private medicine are providing a resource for making 'good citizens'. At the same time they are practicing what I have called 'reverse traffic', that is they are circumventing anti-trafficking laws against the exploitation of people for bits of bodies (Nahman, 2011). This reverse traffic involves bringing eggs to Israel rather than having people travel abroad and is a growing practice in Europe. This is the first in-depth account of this practice and has implications for reproductive travel, or what I call 'repro-migrations'.

Race, migration and multiculturalism

> Pluralism and Segregation: As a multi-ethnic, multicultural, multi-religious, and multi-lingual society, Israel has a high level of informal segregation patterns. While groups are not separated by official policy, a number of different sectors within the society are somewhat segregated and maintain their strong cultural, religious, ideological, and/or ethnic identity.
>
> However, despite a fairly high degree of social cleavage, some economic disparities and an often overheated political life, the society is relatively balanced and stable. The low level of social conflict between the different groups, notwithstanding an inherent potential for social unrest, can be attributed to the country's judicial and political systems, which represent strict legal and civic equality.

> Thus, Israel is not a melting pot society, but rather more of a mosaic made up of different population groups coexisting in the framework of a democratic state. (Ministry of Foreign Affairs, Israel, 2010)

Race and racism (with the concomitant practices of creation of 'difference' and promotion of eugenics) play a central role in the making of nation and state, as part of an imperialist project (Goldberg, 2002; Lentin and Lentin, 2008). The Israeli state was founded on the related myths of national diversity and harmony (difference and mixture), as well as blood/kinship and religion (Jewishness). It was also founded on the idea of the need for a Jewish ethnic state to combat (or retrospectively prevent?) the Holocaust largely perpetrated against European Jews. For that purpose, non-European Jews were recruited to join the Zionist project and shift from a position of relative comfort and wealth in their homelands of origin to one of misery, poverty and the dual practices of national othering, minoritization and fetishizfation. Built into the Israeli national project of making a safe haven for 'the Jews', was always another parallel one of making it a space of European enlightenment. This project of enlightenment assumed that those Jews who came from Muslim, African or Arab countries did not have the level of civilization to participate in this project (unless they had huge capital resources). This is racist state ideology to the core and persists today in quotidian forms of love and hate that make race politics (Ahmed, 2004).

The biopolitics of the Israeli state involve racializing and minoritizing people whilst transmitting a sense of 'diversity' and tolerance for difference, like many liberal democracies. After a couple of decades of promoting the ethos of *mizug galuyot* (assimilation; a blending of exiles or melting-pot ideology), the Israeli state shifted towards an ethos of diversity. More recently, however, even the semblance of an appreciation of difference and multiplicity are being legislated away, with the 'Loyalty' to the State of Israel as a Jewish State Law, requiring all non-Jews wishing to gain Israeli citizenship to declare loyalty to the State of Israel or the Nationality and Entry into Israel Law, barring non-Jews from the West Bank or Gaza from living with their spouses within Israeli borders. Or, finally, the Egg Donation Law preventing cross-religion egg donation (this law doesn't have jurisdiction over the import of eggs). All of these demonstrate the shutting down of a notion of multiculturalism. But even prior to this, there was no real multiculturalism for Jews self-identifying as Mizrahim (Jews who self-identify as originating from Muslim, Arab, North African and Asian countries). These forms of racism sparked the rise of the Israeli Black Panther movement of the 1970s, and groups such as *Ha-Keshet Ha-Democratit* (The Democratic Arch), or *Achoti* (My Sister), all of whom have fought against the racism of the Israeli state. These groups have called for an understanding of the fundamental links between Palestinians and Arab Jews, against a state that is viewed as an offshoot of European imperialism (Alcalay, 1993; Shohat, 1989; Lavie, 1996, 2011a, 2011b; Yuval-Davis, 1997).

The State of Israel positions itself as the enlightened Western society within a barbaric Middle East, and being a 'legitimate citizen' who has not only formal rights but actual substantive rights depends in part on being designated as Western or European. In the past this has meant 'civilizing' the non-European Jewish population and excluding the Palestinian one.

This 'state project' became realized through many social projects, and Israeli sociologists and anthropologists were often central to this. Many Israeli anthropologists study particular ethnic groupings within Israel in ways that further reified these groups (Lavie, 1996; Dominguez, 1989), playing into the Israeli state's marginalization of those groupings. Dominguez outlines crucial points made by an Israeli anthropologist on this subject in a paper presented at a conference in 1980 (Bernstein, 1980: 262). In a series of five main points, Bernstein summarizes the ways in which sociologists literally follow the government's ideology and create sociological conceptions for these – so, for instance, if the government treated new immigrants as 'pre-citizens', 'passive dependants' and 'as children', the sociologists followed suit by introducing terms such as 'desocialization' and 'resocialization'. This is crucial to understanding the academic (and social) apartheid (Lavie, 2011b) that exists within Israel.

What makes some mixtures more worthy of life – more 'vital' than others? In the study of Israeli identities and 'society', ethnicity has been used as a social category to put people into as part of the civilizing missions of the government. In later times, to study the exotic other of the Ashkenazi elite was a privileged purview of the dominant national subject. Anthropology and sociology have been guilty of presenting the 'Mizrahim' as objects of study and rarely as speaking subjects with an ability to enact critique (Motzafi-Haller, 2004: 178). Now Mizrahi feminists have been calling for an attention to *Mizrahiyut*, marginalized subjects and in this case notions of mixture, not in order to objectify and speak about these from a distance but to see what constructions of Israeliness can be most transformative or have the potential to do Israeliness differently (Lavie, 2011b; Motzafi-Haller, 2004).

Israeli IVF cannot be examined outside the history and contemporary discursive practices of state Zionism. Israeli IVF is structured by the history of the Israeli state and by contemporary discursive practices of Zionism. Israel is a settler colonial state encouraging the reproduction of Jews where women reproduce the nation and are women recruited into the collectivity for that purpose, although not all women equally (Amir and Benjamin, 1997; Berkovitch, 1997; Kahn, 2000; Kanaaneh, 2002; Portugese, 1998; Yuval-Davis, 1997; Weiss, 2002). Zionism fights ideologically and culturally to remain a viable vision of the world for its citizens. An appreciation of the ways in which the cultural heritage of European Jews reproduces itself and remains dominant is central here. An Ashkenazi identity, that is a European identity, confers on one a right to legitimate citizenship in a manner different to any other in Israel. The fact that state Zionism officially declares that it is a multicultural society is consistently betrayed by the social and cultural realities of the privileging of Europeanness (Lavie, 1996; Shohat, 1989; Alcalay, 1993).

These practices of race and racism are seen throughout *Extractions*, through recipients' selection of donor traits, through practices of erasure of non-dominant subjects in clinics or the conflation of Arab Jew with non-Jew. The majority of my research participants were non-hegemonic Israelis. Their ambivalences about Arab eggs are seen as moments of possible resistance to the state and their own positioning on its real and imagined borders.

Extractions 4: cyborg feminist writing

Here is a mapping of a relation: the first Israeli egg donation occurred in 1984, just three years before the First Palestinian Intifada. This story of transnational, privatized, egg donation in Israel cannot be told without the story of the occupation and Intifada, which infused the daily activities and interviews during the research for this book. These are the 'partial connections' (Strathern, 1991) that *Extractions* maps. Not only were the connections between war and reproduction manifested in images and texts that I observed for nine months in public and private Israeli ART clinics in the year 2002, but they are also deliberately brought into the ethnographic storytelling as evidence of how historical time and location shapes ethnographic rendering, and to question how to tell a story about reproductive technologies at a time of war and occupation. Anthropologists have shown how reproduction can be stratified and racialized (Colen, 1995; Ginsburg and Rapp, 1995). Here, stories from clinics, from streets and from homes are used to discuss impurity, immobility across bodily, geopolitical and disciplinary borders. These boundaries, like cells in a body are dense – these are not lines but many layers, overlapping and overlaid. The antiseptic worlds of both IVF practice and ethnographic description are deliberately disrupted through a situated form of writing that enacts a contestatory politics. These are based on the understandings of the research participants and how they viewed their worlds as well as on my own interpretation.

This book diffracts the discursive practices of extraction, exchange and transfer of human oocytes between differently situated women, evoking a figuratively synecdochical relationship between the egg and the nation. 'Diffraction is a mapping of interference, not of replication, reflection or reproduction' (Haraway, 2004: 70). Diffraction is the way in which light waves are changed when put through a prism – the light wave is changed or interfered with. We already know through anthropological and feminist critique of representational practices that we always affect those we are studying and that the author creates as much as she describes. Diffraction is about writing in a way that does not merely describe, replicate or reflect – rather, it makes something new. It is about placing politics and power at the centre of the social scientific practice of writing. When we go out there to study something, we are not merely representing it – we are making interferences.

Anthropology has a history of being uncomfortable with intervening in the cultures that we study. And with good reason. Part of our disciplinary history comes from assisting in colonial administration. Anthropology's legacy of cultural relativism, from Ruth Benedict, Boas and onwards, represents a strong disciplinary bias against intervening in other cultures. The ethos is that cultures should be taken on their own terms and not judged. To intervene on the basis of one's own ethics or cultural precepts would be to presume those to be superior. Also, to intervene on behalf of others can be seen as patronizing. Yet there are times when intervention seems necessary despite the potential of being seen to be 'culturally imperialist'. Indeed, it has been argued that there is a long history of value-laden work in anthropology, which histories of anthropology have only partially represented and that understanding the world must not be a substitute for being committed to improving it (Rylko-Bauer, 2006: 183). So the synecdochical or part-for-whole connections that this book maps between eggs and the nation are made by me.

An early form of Zionist thinking was that:

> all the laws and ordinances, all the blessings and curses of the Law of Moses have but one unvarying object: the well-being of the nation

> as a whole in the land of its inheritance – the happiness of the individual is not regarded. The individual Israelite is treated as standing to the people of Israel [as] in the relation of a single limb to the whole body: the actions of the individual have their reward in the good of the community. (Ahad Ha'Am, 1889)

In terms of even the 'biological' body then, the individual *is* the collective. Importantly, the individual that represents the national is Jewish, Ashkenazi, able-bodied. In Judaism, one is supposed to imagine oneself as standing in for the collective. Synecdoche then is not just an indigenous practice to 'nationalism' but to Judaism itself.

Anthropologists have long questioned their practice of 'making cultures' through their writing (Clifford and Marcus, 1986; Geertz, 1973), and feminists were central to these discussions (Strathern, 1988; Visweswaran, 1994; Behar, 2008; Wolf, 1992). Hayden White's critique of representational practices within historiography specifies that the way in which conceptual and representational practices are enacted is political (White, 1987). Lila Abu-Lughod's (1993) critique of the narrow ways of representing 'woman' in the Middle East through concepts of 'shame', 'honour' and so on are both emblematic of the kinds of critique that this book is trying to present. I have deliberately chosen to highlight war, violence, multiplicity of women's voices and the shifting nature of borders and boundaries. This is in order to undo the very ideological framing of Israeli reproduction as something under threat, something that is merely a representation of the state's own aims. Haraway's 'Cyborg Manifesto' (1991) enjoined feminists to rethink the relationships between theory and practice, the body and technology across differently situated but similarly motivated feminist positions. According to Haraway, 'cyborg thinking' is a way of doing feminism across disciplinary boundaries in order to undo systems and ways of seeing the world. Both the feminist philosopher Rosi Braidotti (1994) and Haraway call for a proliferation of feminist figurations that subvert 'old ways' of viewing what happens to/in/through women's bodies. In Haraway's case, the focus is mainly on the relations between science or technoscience and the materiality of bodies. The synecdoches I make in this book are based on my interpretation of what I observed and what people said to me during the course of the research.

One way of figuring this relationship between the stories about recipients and the story of the state might be through an understanding of the constant 'ricochet effects' (Hayden, 1995; Strathern, 1995) that occur in meanings and practices as they move between the domain of the body and the domain of the nation, simultaneously producing both. This is a figuration in which the body does not literally stand in for the nation; this is not a direct synecdochic relationship, as in Mary Douglas's notion of pollution and broader communities (1966). But materially and metaphorically there are synecdochic resonances. Elsewhere, I have called the process of weaving together these moments in the research process 'biosocial ethnographic moments' (Nahman, 2007), which indicates that there are moments in the practice and process of ovum extraction and exchange that articulate relationships between the state and reproduction. Except that this relationship is neither directly governmental in the Foucaultian sense of biopolitics, nor is it agentic in Paul Rabinow's sense of the biosocial. The kind of relationship that these situations map out is one of synecdochic ricochet (ibid.). In stories about egg donation the individual woman's body may be figured as saying something about a larger national set of issues.

Yet, crucially, *I* am reading the resonances between the nation, the state, the war and the body as a way of contesting representations of women and the nation in Israel. My juxtaposition of the issues confronting the nation and those narrated to me by people involved in Israeli ova donation is one example of a mapping out of a similar story told in different but resonant spheres, following Marilyn Strathern's (1988, 1991) illustration of the productive nature of such a project. There are other instances where research participants themselves directly linked the need to reproduce individually with the need to reproduce the nation, but these are not the main focus here.

Asking whether eggs can stand in for the nation is a way of creating a story about race, gender and politics in Israel. This involves an examination of claims made by research participants and has an ethic of describing what I see as the most crucial links between the ways in which eggs cross bodily and national borders while other kinds of mobilities are more constrained (or even prohibited). Without essentializing the kinds of bodies in question (for example, I do not assert that all 'women' reproduce 'families' or 'nations'), I invoke synecdoche in order to assert that subjects are materialized through the relations that are produced by exchanges, extractions, manipulations and legal debates about bodies and body parts. This is a situated and particular account of cultural practices, the story itself is constructed, as they all are.

Extractions puts stories from the clinics and interviews alongside stories about the Intifada as I experienced it and as documented at the time by the Israeli human rights organization, B'Tselem. The juxtaposition of moments of crisis that are not directly linked to one another is done in the hope of contesting how stories about Israeli nationalism and women's reproduction for the nation are told, making a new story. Mapping out these 'extractions' might be one way to do so.

The research transforms the ethnographer

> [I]t is not so much that I have written this book, as that it has written me. Or rather, 'we' have 'intra-actively' written each other [...]. Which is not to deny my own agency (as it were) but to call into question the nature of agency and its presumed localization within individuals [...]. (Barad, 2007: ix–x)

'Michal is the only one in this panel who is not living and writing from within Israel,' an eminent feminist anthropologist said accusingly. So what? I was the only one on the panel overtly critiquing the Israeli state project – bringing up what was to me now an obvious contradiction in terms: a Jewish and democratic state; criticizing the culture of crisis and victimhood that pervades Israeli society and enables participation in a national project that actually makes victims of Jews (if we look at numbers of people dying in road accidents, bombings, shootings and disease, Israel is the most unsafe place to be a Jew in the world right now). There were other critics of Zionism on that panel – but they spoke about other topics and so were less obvious targets. At the time I defended my right to speak as someone who had grown up amongst Israelis (in Canada), and also on the basis that I had only narrowly escaped three suicide bombings whilst conducting my research in Israel. But this recourse to having been 'vicitimized' and having Israelis for parents, friends, teachers and Hebrew as my mother tongue seems now to be beside the point. We have had this discussion in feminist writing, medical anthropology and elsewhere: can only women speak about 'women's issues' or can only patients speak about illness?

I was not always a critic of Zionism. The research transformed me.

In fact, I was a committed but questioning Zionist until I conducted the research on which this book is based. I was raised to think of myself as an Israeli Jew. I have a name that is barely pronounceable to non-Israelis – even the other kids at my all-Jewish schools could hardly say it. I would celebrate Israeli Independence Day in school every year. At my high school when we sang the Israeli national anthem *Ha-Tikva* (*The Hope*) every morning – I sang the loudest and proudest. I was vehemently opposed to the PLO and believed strongly that they were a 'bunch of terrorists' who wanted to steal Jewish lives, just like the Nazis and the Cossacks before them. Right there, collapsed into one was a victimhood, victimizer, victimhood narrative. Indeed, I was raised on a whole host of visions of attackers of the Jewish people: the Greeks who destroyed the Temple in Jerusalem and desecrated it with pig's blood; the blood libels of the Middle Ages where Jews were killed for having survived plagues (because if they survived they must have caused them). On Holocaust Memorial Day as young children we were taken to the basement of the school, with lights dimmed and candles lit to look at photos and sculptures of representations of the demise of our people. One image of a boy behind barbed wire was actually now our school principal. A sculpture carved out of smooth greyish stone was of dead naked bodies piled into a heap. This terrifying spectacle worked, it created scared, vigilant children out of us. We were afraid even to glance at a church or a cross in the street and would avert our eyes if we passed by one for fear of dying or being converted. It was clear to us that the world was out to get us.

But there was another side to this. We were eternal victims who could fight back. We had a powerful army that we were told was like a David to the Goliath of the surrounding Arab states. Should a mass wave of anti-Semitism hit Canada, we could count on the Israeli army to protect us. I grew up dreaming of serving in the Israeli army. I had an Israeli passport from the age of 12. At that age I spent a summer in Israel and attended summer camp where I learned to fire a rifle and a crossbow.

Going to live in Israel again in 2002 seemed like a different thing altogether. I had already started to read critiques of Israeli politics and was very confused about what I felt or thought. I most certainly believed that the State of Israel had a right to exist as a Jewish state (something that after living there I was not so sure about anymore), but had started to wonder about what had actually happened in 1948 when the State of Israel was founded. I had begun to hear about the *Naqba*, what Palestinians call 'The Catastrophe', which is this same founding.[5] It was living there that transformed my thinking most profoundly. Witnessing the Second Intifada through the eyes of the Israeli media, everyday interaction with Israelis, narrowly escaping bombings. The media reports were extremely one-sided; it was very challenging to get any news of what was happening to Palestinians in their towns and homes. Yet it was palpable that something was happening because of the intensified military presence on the streets and in the skies. Nighttime bombings could be heard, many more military helicopters than usual could be seen. Large convoys of armoured personnel carriers were to be seen on the highways.

This made me start to seek out other perspectives – it seemed all too clear to me that there were more angles to the story than what then Prime Minister Ariel Sharon wanted us to think: 'We are in a war over home and we must unseat the terror infrastructure [*tashtit ha-teror*].' I befriended one of the founders of Checkpoint Watch (then a small group of women activists whose aim it was to prevent inhumane practices of Israeli soldiers towards Palestinians at checkpoints), an anthropologist who provided sources and resources for another view. I attended a Gush Shalom (an Israeli peace organization) protest at Rabin Square in Tel Aviv, where we argued that Arafat was indeed a partner for peace. Standing there among approximately fifteen other protesters was inspiring – despite being spat at and having eggs thrown at us by young children. Later I attended a Women's Refusal Tent protest in Tel Aviv – a week-long event that drew together women from different Israeli backgrounds all against the occupation of Palestine. These were Palestinian women, Ashkenazi women and Mizrahi women performing a critique on the kind of activism that excludes non-normative Israelis (the major mainstream Israeli left consists of Ashkenazi – read white – middle-class Israelis, many of whom look down upon their non-Ashkenazi counterparts; there are long histories of divisions within the Israeli left). I attended a conference of Mizrahi feminists where they berated others for exclusionary, racist behaviours and attitudes.

Becoming part of the Israeli left was transformative for this project and for me in general. It explains why this book had to be written in the way in which it has been. Not only was this my experience of the research, but it is also demonstrative of how the critique itself emerged. Images of graffiti outside clinics, drawings and photos inside clinics, excerpts from fieldnotes detailing the progress of the attacks on Palestinians and Israelis show this transformation.

In all of the chapters I offer excerpts and extracts from my fieldnotes. Many of these present my perspective on the Israeli occupation of Palestine at the time and its effects on everyday life in Israel. In order to supplement these with information about the experiences of Palestinians at the time, I have used information from the Israeli human rights organization, B'Tselem. All of these appear interspersed with and interrupting the main ethnographic text.

Between March and April 2002, my observation of what was happening all around me turned me from a questioning Zionist into a critic of the State of Israel. My early journal entries of March refer to the 'territories' a Zionist euphemism, but by April I was thinking of them unequivocally as the Occupied Palestinian Territories. In earlier notes,

I refer to a suicide bombing as a '*piggu'a*', a word for attack derived from the word 'injury'. By April, I had started calling them 'bombings' and noting the actions of the IDF against peaceful observers. These excerpts demonstrate how the research transforms the researcher as well as the intimate intertwining of reproduction, violence and occupation.

Overview of the chapters

The voices of Romanian women who decide to sell their eggs to a private egg donation clinic are the focus of Chapter 2. Their decisions, informed and uninformed, to sell their eggs are made for different reasons, usually monetary gain is the main driver. But often it is a desire to do something good, or fun that drives them. Taken in conjunction with the repressive history of reproduction in Romania (illegality of abortion, adoption 'tourism', aggressive pronatalism), these choices make Romanian women into a new kind of person. They 'sell their eggs to get a better life', as one woman said, and to gain a better standing for themselves as women. This private Israeli clinic running an oocyte extraction clinic in Bucharest, Romania, is a site where the bloody and immobilizing outcomes of capitalism's incursion into bodies can be seen most starkly.

The ways that Israeli nationalism happens transnationally through private consumption practices coupled with religious and messianic hopes for Israeliness are explored in Chapter 3. This chapter examines egg donation at a private IVF clinic. Company logos and brands set up the clinic as a space of consumer comfort and ease, whilst narratives of the people who run the clinic maintain it as a space where national aspirations can be achieved. The nation is materialized as elastic and inelastic at the same time, the clinic a space where consumer capitalism, religion, private medicine and Zionism mix. To some extent, the way in which transnational ova donation happens privately, transnationally and across religions, disrupts any straightforward idea that the state is managing the population. It disrupts a formulation of the notion of an 'Israeli woman' guarding borders and the discourse that it is provided freely and equally to all of its citizens.

Israeli borders and national becomings are examined through the voices of Jewish and non-Jewish Israeli egg recipients in Chapter 4. The ways in which becoming an egg recipient involves wanting to become invisible, and not wanting to travel, indicate the problematic nature of the notion of reproductive 'tourism'. Instead, 'repro-migration' is suggested as a way of thinking about transnational reproduction, histories of migration to Israel and present-day racializations together, in terms of how Israeli egg recipients view themselves and their reproductive travel. This notion of repro-migration is followed in the second half of the chapter with an account of the racializing of Israeli egg donation. The desirability of whiteness is examined through the recipients' selection of desirable donor/egg/baby traits. The ambivalence of 'Third World' Jewish Israelis around accepting Arab eggs emphasizes how they have always been positioned on the border of acceptability in Israel. The ways in which they both resist and reaffirm racialized Jewish–Arab boundaries is explored through their own choices.

Chapter 5 explores the claim that Israel is a state on the vanguard of scientific and technological research and development (R&D). This is seen as a nationalistic discourse intimately tied to the discourse of the 'need to protect national borders and bodies', and the disposability of Israeli lives. This chapter provides a 'thick description' of oocyte extraction, fertilization and implantation in a state-funded IVF clinic. Border-making, through violent metaphors, is deliberately and unavoidably juxtaposed with border-making and violence that was happening all around us during the course of the research.

The research chapters conclude with Chapter 6 on crisis, Israeli nationalism and egg donation. Banal nationalism (Billig, 1995) is said to occur in mundane quotidian ways – in contrast to extreme forms of nationalism. However, in Israel, extreme forms of nationalism *are* banal. Chapter 6 theorizes a relationship between women, body parts and the nation. Through an examination of accounts of 'the episode that exploded', an egg theft at a private clinic in Israel in the year 2000, this chapter argues that women's bodies may be thought of as vehicles for narrating national victimhood. The perceived crisis of a shortage of eggs resulting from this theft of eggs is described and it is suggested that something else is taking place: a reification of national identities, borders and bodies.

Oocytes are the largest cells in the human body. They are vessels and vectors of volumes of possibilities and stories. Like water, these stories shift, accumulate, dissipate, crash, solidify. These are global possibilities and stories of inequality, and how capitalism and new kinds of liberalism make the spaces between differently situated people wider. This is about the racial and class desirability of eggs. The ethical questions and moral reckonings in global ova exchanges that bind differently situated people in relationships of love, gratitude and monetary gain also serve to widen the gap between them. Transnational kinships, either of a fictive kind or genetic, expose the glaring chasm that forms from the economic (read moral, ethical and bioethical) systems within which they were made. Despite the very real, painful or joyous effects of oocyte extraction and transfer, telling the story of these practices makes and remakes eggs into something else: an object of ethnographic knowledge, a contested sphere. So there are various ways to tell the story of Israeli egg donation. It's about how you choose to map them. These choices are crucial.

2
TheoristSellers

The Blood in Aspirator I

It's 7:30 a.m., 9 September 2002. The surgery room where oocyte extractions are performed is on the top floor of a purpose-built oocyte extraction clinic, in a converted house, near the centre of Bucharest. It is in an inconspicuous building that appeared, from the outside, to be a small three-storey house. Externally, it blends in well with the run-down neighbourhood. It appeared to have been recently renovated; a clean and well-equipped building. At the entrance level is a secured, locked gate, guarded 24 hours a day by a uniformed man. Opposite is a derelict house, with a front yard populated by people, domestic objects and a dog. The main floor has a large waiting room, with about ten chairs. These face a doctor's office/kitchen, where medicine is kept in a refrigerator together with food. Next to this is a nurse's office. Behind the waiting room is an enclosed, paved yard. On the first floor are three bedrooms, one belonging to Dr Shmuel,[1] another to Sasha (the embryologist from Israel) and an extra room in which staff occasionally rest or sleep. Adjacent to one of the bedrooms is a tiled, grey bathroom and shower. The next floor above is dedicated to the surgery room, recovery room, lab and pre-surgery shower room.

Inside the surgery room (see Figure 2.1), all the machines are ready. There is a hospital bed, with foot rests, a trolley for the anaesthetics, oxygen canister, a low cabinet with drawers and various utensils and liquids along one wall. Then there is the ultrasound transducer, a microphone-shaped implement, with a long needle that slides through attached loops and runs parallel to the transducer. Finally, there is the 'aspirator'. A machine connected via clear tubing to the needle. This machine creates suction for extracting follicular fluid from ovaries. This fluid is

Figure 2.1 Surgery-room bed, monitor and aspirator, Bucharest clinic
Source: The author.

meant to contain oocytes, small human eggs. Many of the machines, furniture, tables and pharmaceuticals were transported from Israel to Romania. Many bear logos of Israeli manufacturers in Hebrew writing.

All the machines are in place and ready, the people begin to enact the choreography that is intended to bring about the 'aspiration' (or removal) of oocytes from a woman's body. A few minutes into the first operation, the choreography changes drastically because the aspirator becomes clogged with blood. There is so much blood stuck in the machine that it fails to produce a suction that would normally extract eggs. Telephone calls are made using mobile phones. In these calls information is passed along that results in the humans needing to remove the clear tubing from the aspirator and insert a syringe on the end of it that provides the suction to extract eggs. It takes some minutes for this to work, during which time more white fluid is injected into the patient's arm to make her sleep.

The blood in Aspirator II: inserting emotion and the ethnographer

The first woman was called into surgery. She had red hair and was very slim. I followed her into the surgery room. The lights were bright and the room was cold. This young woman was very talkative and spoke quickly in Romanian to the staff. Iulia, the physician, joked with her. The anaesthetist, Maria, joined in while she busily injected a white liquid into the seller's arm through an intravenous drip. I noticed that there was no pulse monitor, and made a note to ask the anaesthetist about this later. Soon, Natalia came in to assist with the operation. She covered the transducer (ultrasound probe) with a large latex cover.

I saw Sasha in the laboratory preparing various mediums and implements. A few moments later I heard him on the phone with Dr Shmuel who was still in Israel. It seemed as though Dr Shmuel had asked him if there had been any *takkalot*, hitches. 'The only *takkalot*', Sasha joked, 'are that it's too early to be awake and no one has made me any tea.'

The soon-to-be-seller was now asleep and the lights were turned down. There was a beeping sound coming from the machine that sucks the oocytes. Natalia and Iulia turned their attention away from the young woman on the bed and focused on the machine. At one point they turned the lights on in order to be able to see well. Then, three of them (the medical assistant had joined in) were focused on the machine while the anaesthetist and I watched. The prospective seller was still unconscious, her legs up in stirrups. They had already disinfected her with iodine and isotonic saline solution, but the machine for aspirating the oocytes was not working. Finally they called Sasha to assist with the machine. He took over.

Ten minutes passed. The seller was still lying asleep, no one was checking her breathing now, they were all focused on the machine. She awoke and no one except me noticed. When they did notice, they told her, gently, to lie still. At one point, she sat up and grabbed Iulia weakly by the arm. Iulia calmed her by asking her to '*stai acolo puţin*', 'stay there a little'.

The confusion over the machine persisted and they rang Dr Shmuel in Israel for help. The medical assistant, Clara, went over to the seller, caressed her head, smiled and talked to her quietly. She continued to hold the seller's hand. We waited and waited. Finally, Dr Shmuel phoned Sasha. Despite the fact that Dr Shmuel is a native Romanian speaker and Sasha did not speak much of the language, all communication from Dr Shmuel went via Sasha. Sasha explained that he thought *she-yesh dam ba-mash'eva*', 'that there is blood in the pump'. Blood had become stuck and had clogged up the aspirator. Dr Shmuel explained to Sasha an alternative method of aspirating the oocytes. Sasha got off the phone with Dr Shmuel and asked Natalia to bring him a syringe, which she promptly did. Sasha took this and attached it to one end of the long extraction tube. He then explained to Iulia what she must do. The entire staff was giggling and cracking jokes. Iulia said that she did not understand something (I did not catch what it was). Then, Sasha seemed to have cleared up the misunderstanding. The syringe was given to Nadia, the medical assistant who was trying to figure out what to do with the syringe, in order to create the vacuum that would suck up the follicular fluid containing oocyte. When all was cleared up, Sasha took out the aspirator to dry it.

The woman on the operating table was given more anaesthetic and the operation continued. One assistant emptied the air out of the pump and pulled on the syringe in order to suck out the oocytes and fluid that Dr Iulia had extracted. She did so by poking the needle at the follicles, which she viewed on the ultrasound monitor.

Despite the initial problem of blood being stuck inside the aspirator, there were more eggs being extracted than I had ever seen before (at this point I had been observing oocyte extraction for eight months). There was blood in the test tubes. I reminded myself that blood does occasionally come out in extractions. I was told in Israel that this tends to happen when the physician mistakes a blood vessel for a follicle containing oocytes. The seller jumped slightly and tried to move and the anaesthetist held her arm. When the ultrasound stick came out of her body, a large chunk of something bloody came out with it. There was a lot of blood on the paper sheet beneath the seller. On seeing this, I struggled to maintain composure.

Natalia opened the next round of test tubes without changing her gloves. She rinsed the needle tube in the sink by pushing water through it with the syringe. The seller was still sleeping. A bed from the recovery room was wheeled in. A medical assistant entered (I noticed at this point that they were all wearing the same hospital garb as they do in the clinic in Tel Aviv, with the same logo – see Chapter 3). This medical assistant cleaned the young woman a bit with some cotton. They unbound her arms and legs and attached her to a mobile drip. She was still asleep when they rolled her onto the stretcher. They covered her with a sheet and wheeled her out of the operating room.

A new sheet of paper was rolled onto the surgical bed, the anaesthetist wiped the bloodied end of the bed with some paper and the next seller, smiling and tanned, was brought in and laid down on the bed. One of the medical assistants cleared out the previous oocyte seller's shoes. Iulia put on a new pair of rubber gloves. Natalia prepared some iodine, while Maria put her patient to sleep. Natalia opened the package containing the transducer cover and asked the seller how she was feeling.

The seller, still smiling, said, '*Bine*,' 'Good.' The voices in the operating room were not hushed, and at one point Iulia shouted to another staff member who was outside in the corridor. The patient, who now seemed to be unconscious, moved a little as she was being disinfected. The anaesthetist pushed further on the injection in the woman's arm. The operation continued without a hitch.

After observing a few more extractions, I returned to the bedroom on the second floor of the clinic where I was staying. I was desperate for a break. The visions of extractions, of the aspirator breaking down, had left me drained and upset. Back in my room I began to sob.[2]

The emotional effects of such research on the ethnographer cannot be overstated. I am still left with the sense of anxiety that I had whilst standing in the surgery room watching oocytes being extracted, seeing what seemed to me inordinate amounts of blood coming out of a woman's body. The walls of the clinic were covered in a variety of deeply coded images. I was unable to watch and am unable to write about all this with a dispassionate eye. Two years previously, I had been asked to 'donate' my own eggs to a woman in exchange for significant financial compensation (Chapter 4). That I did not go through with it meant that in the space of the clinic I too was haunted by my own potential donor/seller status. The sense that 'that could have been me' on the operating table, was quickly followed by the understanding that 'that never could have been me', given my upbringing in a well-to-do Canadian suburb. It never could have been me because of the global economic order that offers substantially larger sums of money to women with my social, religious and geographic positioning. This is not about guilt, but positioning. The global economy of eggs means that Jewish Canadian university graduates get more money than Romanian women.[3] This shapes the story/stories that I have told here and throughout the book.

Explaining 'it'

This chapter explores the practices of transnational ovum extraction in an Israeli clinic in Romania. It examines the voices of Romanian egg sellers themselves as a source of theoretical and political direction for transnational feminists who try to think about responses to reproductive technologies. That the women who sell their eggs are also 'theorists of global reproduction' is the central claim of this chapter. Their 'theory' is that they sell their eggs to get a better life. This is an embodied pragmatics of inter-cumulating. Or a method of coping with living under capitalism, on its margins. The entire first half of the chapter is an attempt to explicate their 'theory' through a thick descriptive analysis of the practices of ovum extraction or 'aspiration' at this global clinic, whilst in the latter half the women spell it out themselves through interview excerpts. Little has been heard from the women in Eastern Europe themselves who have opted for selling or trading their oocytes. The sellers were interviewed in September 2002 in the Bucharest clinic of an Israeli company that specializes in transnational oocyte trade and has IVF clinics in Israel, the United States and Western Europe (Chapter 3 examines the practices of that clinic more closely).

> At the time of these interviews, the Al Aqsa Intifada was ongoing in Israel. The Israeli Human Rights organization B'Tselem cites several cases of Palestinians being used by the IDF as human shields during the month of September 2002 (B'Tselem, [2011]).

It is plain that this is not the kind of account that can be told from a distanced stance. It is also urgent that this is not seen as a sensational story. It is a story about 'exploitation' and 'choices', but one that must be taken together with a wider understanding of the kinds of labour (reproductive and otherwise) that are possible and quotidian for people in the context of Romania. So, fundamentally, it is a critique of the way in which capitalism mines workers and their bodily resources in 'marginal' spaces. Here we are at the borders of Europe, which is at once the centre of capitalist markets extraction of value from workers and their bodies, at the same time it is at the outer edges of inclusion into liberal democratic rights.

As I began to argue in Chapter 1, egg donation is not one 'thing' or 'process'. It is not an objectification or reification, which shuts down the ability to interpret and describe what happens when you take eggs from one woman's body and put them into another's. To turn egg donation into an 'it' is to forget how, in globalization and science, there are many ways of doing something with many possible outcomes. 'It' happens differently in different places, and times, in as many ways as it is 'the same' in these places. Egg donation in Israel is very different from egg donation in Europe, the USA, India and in this case, Romania. Seeing such multiplicity can help expand the interpretation of what counts as 'trafficking', 'women's reproductive labour' and 'exploitation' in global oocyte extraction, exchange and implantation (Dickenson, 2007; Waldby and Cooper, 2010; Widdows, 2009). Without attempting to answer these questions directly, this chapter asks: What is the temporal and geopolitical specificity of transnational oocyte extraction and 'traffic'? One way for such issues to be historicized and located is through an ethnographic anchor. The descriptions of oocyte extractions, observations in clinics, accounts of interviews and waiting-room discussions in this chapter set the stage for such anchoring.[4]

Ceausescu's anti-abortion laws, poverty, unemployment and women demanding abortions can signal the complicated and ambivalent relations that the women I spoke to had with their bodies, global economies and biomedicine. Some of the egg-selling stories below bear echoes of the denial of access to safe technological means to prevent reproduction, it occasionally seemed as though a remnant of that repressive state past resided in the sellers' desire to sell oocytes as an act of resistance against that past. This is illustrated below in an extract from an interview with one woman whose logic for selling her oocyte is partly about doing what she wants with her own body. The politics of reproduction in Romania have been well documented in relation to abortion, adoption and coercion (Kligman, 1998; Gal and Kligman, 2000). In the name of the 'nation' Ceausescu aimed to prevent abortions and made it illegal (unless a woman had at least five children). Concomitantly, illegal abortions rose, as did the numbers of orphaned children and maternal deaths (over 500 in 1989, with an immediate decrease to 100 in 1990 after Ceausescu was removed from power (World Health Organization (WHO), 1990, cited in Hord et al., 1991)). This was an ironic outcome of state pronatalism.

Another outcome was what Gail Kilgman called 'adoption tourism', the purchase of a child by non-Romanians through an agent. Kligman notes that poor and ethnically minoritized Romanians received money in this way to 'modernize' their lives by buying radios, laying down flooring or purchasing furniture (Kligman, 1992). Post-1989, Romanian feminists and NGOs worked to educate women about contraception and to attempt to introduce teaching on 'family planning' in schools, but they encountered resistance from within the government. Abortion remained the contraception of choice for many Romanian women.

The oocyte trade in Romania began in the 1990s. There were no regulations against paying women for their ova. It seems important to contextualize this in the history of abortions and adoption 'tourism'. 'Biological citizens' (Novas and Rose, 2005) responding to extended state control over their bodies, women were now doing reproduction differently. Still secretive and seen as 'shameful' – the comparison to sex work no doubt entered their minds – there was nonetheless a different quality to egg donation when seen in person and heard about from the women themselves. No men or states had coerced them. The sheer availability of the route to cash led them to it. Female friends led them to it. Romanian egg selling signals both a continuation of the conscription of women's bodies (by and for science and business) and a break with paternalistic control over women's bodies. This argument indicates that there may be a significant tension between the priorities of Western Euro-American feminists and the women whom I spoke to.

The oocyte selling stories also mark the lack of real material improvement in the lives of many Romanians since Ceausescu's time. Romania lost more than a quarter of its labour force after 1989, with about 2.5 million Romanians migrating abroad. This led to an increase in labour shortages, especially in the arenas of construction (where there was a 50 per cent shortage), healthcare, IT and tourism. On the one hand, these shortages led to a slight wage increase. In Romania the minimum legal wage in the early 2000s was about €86 (US$85 in 2002) per month (with annual salaries being approximately 10 per cent of those in the European Union). However these skilled labour shortages had repercussions for workers with respect to terms and conditions. Workers do up to 80 hours of work per week, doubling their full-time shifts. While this is not illegal, it requires the stretching of labour laws that suggest a 48-hour-week. In the Romanian labour force, 12 per cent have multiple full-time jobs. This is the highest percentage in all of Europe. For example, 60 per cent of tourism workers work overtime. Fewer than half of these are actually paid for this overtime (Trif, 2008).

Unpaid overtime is also seen to a large extent in the textile and other labour-intensive industries, where a system of transnational trade operates called '*lohn*', meaning 'wage' in German. The *lohn* system depends on an agreement from the early 1990s, where the Central and Eastern European countries and the EU agreed that companies from the old EU states (the EU15) could export goods (either raw materials or goods in the early stages of manufacture) at low cost and pay tax for the goods returned only on 'value added'. In Romania the *lohn* system was used to export about 80 per cent of textiles, clothing or shoes. This accounted for 35 per cent of all Romanian exports. Companies using this system usually set very high targets that are unreachable in the typical working day. Hence work is completed in unpaid overtime. Such arrangements must be viewed within the unequal power relations that characterize EU relationships among old and new member states (Trif, 2005, 2008).

Despite this, Romanian trade unions are among the strongest in Eastern Europe in terms of their influence on legislation and size. After the fall of Ceausescu in 1989, Romania's economy transformed gradually (rather than through the 'shock therapy' of privatization and trade 'liberalization' installed in countries like Poland). In the early 1990s, the Romanian economy was beset by problems such as falling gross domestic product (GDP), high inflation and low foreign investment and it was one of the weakest economies in the region. Since accession to the European Union and from 2000 to 2007, GDP grew by almost 6 per cent annually. Still the employment rate fell from 63 per cent in 2000 to 59 per cent in 2006 (Eurostat, 2007b).

First evening at the clinic

The chief physician granted me entry to his Romanian clinic to observe over the course of two weeks, after I had spent a few months conducting interviews, and observing procedures and day-to-day activities in his Tel Aviv clinic (see Chapter 3). On a kind of 'repro-migration' of my own, I arranged to meet my good friend Louise in Romania in advance, and we travelled together to see where our grandparents had been born. She stayed with me at the IVF clinic, often providing a sounding board, and as someone to help me deal with the intensely emotional nature of what I was researching. Louise and I arrived back in Bucharest after returning from our trip to Sinaia and Brasov in the evening. We were met at the Piaţa Unirii by Dina, the manager at Universal ART, the clinic set up by Dr Shmuel and his business associate (Figure 2.2).

Dina very kindly walked with us to the clinic that, she told us, was in an area that was 'not so great due to the gypsies who live there'. After showing us around Dina let me know that the 'girls' would be arriving for their final pre-surgical injection with the nurse (36 hours prior to surgery) at approximately 10:30 p.m.[5] I refer to the women that I interviewed as ova sellers. However, in the clinic, the sellers themselves and mostly female clinic staff referred to them as 'girls'. I found this language difficult to adopt myself although the women whom I interviewed were generally young and in their early twenties. They were not 'girls' in the sense of being prepubescent females. On the contrary, they made adult decisions to undergo complex surgical procedures. I often wondered if referring to them as 'girls' made it easier for the clinic staff to relate to the sellers. Staff and sellers were only separated by approximately five to ten years in age. Were they communicating this difference in age? Perhaps this reference to 'girls' made it simpler to treat them as resources. The sellers were not 'girls' in my mind. Coming as it did

Figure 2.2 Piaţa Unirii, Bucharest
Source: The author.

from the clinic staff this appellation, 'girls', seemed patronizing. For example, when the administrators referred to sellers as 'our girls [...]' it seemed possessive. I believe that if they had referred to them as 'women', autonomous, thinking individuals, it would be difficult to possess them (would they have said 'our women [...]'?) Further, when the sellers spoke of themselves in relation to the clinic, for example when they deliberated over being a 'good girl' (producing many of eggs, coming on time for injections), the word 'girl' positioned them as resources. This positioning of the sellers as resources may have been more difficult to do had the staff referred to the sellers as 'women'. This naming practice indexes the familial dimensions of IVF medicine, which involve patronage and paternalism (see Chapter 5). My interpretation of the word 'girls' as a problematic way of referring to sellers may be a problem of translation. After all, among the sellers it seemed that the term 'girls' indicated a kind of sisterhood, a commonality of experience and kinship. Perhaps in Romanian society this word has a different connotation, of which I am not aware.

After helping me settle in my bedroom at the clinic, Dina left the clinic to go home and study for her entry examinations to a Master's degree in genetics. The nurse was due to arrive in three hours' time, when she would administer injections. Once the nurse arrived, I spoke to her about procedures and observed for a while. That evening, and on subsequent days, Dina and the nurse told me about some of the 'intake' procedures that precede the injections and surgeries. Seller recruitment happened by word of mouth. This was done until a large 'pool' of sellers formed. As part of her duties at the clinic, Dina recruits sellers for the clinic. While in 2001 the clinic had only 100 sellers, by September 2002 there were already 300.

The following stage in the recruitment process involved the physician Iulia explaining 'the whole process' to the sellers. The sellers receive an information sheet that outlines these procedures. Dina informed me that 'they have to be very serious girls who will not simply start and stop the procedure in the middle'. Dina explained to the prospective sellers the reasons for seeking oocyte sellers. She would tell them, 'there are women who don't have kids, who can't have kids and it's very bad'.

During the recruitment process, some of the prospective sellers express concern with their own genetic material going to someone else. Dina said:

> [T]hey know eggs are genetic. They know eggs and sperm make a baby, but something new, with a part of them. And I try to explain that they are doing something beautiful because a child is beautiful. Some girls understand and some don't. And I tell them, 'go home and think about it and if you want, come back'. I don't like to lie. (Dina, interview, Romania, 8 September 2002)

After coming to the clinic, having various aspects of the process explained to them and having demonstrated that they have understood what is involved, the prospective sellers sign consent forms and are scheduled in to begin screening and injections. When young women come to be screened for donation they are tested for syphilis, HIV and hepatitis A, B and C. They are also given another blood test to determine if they are allergic to the anaesthesia.

Dina and Iulia would explain the process, outlining the risks and answering questions. Some women whom I interviewed indicated that they felt that the staff members were evasive in response to questions about risk relating to their ability to have children in the future. Dina assured me that every effort was made to ensure that prospective sellers have fully understood the process. They put a special mark on the files of women who seem not to have fully comprehended the process, so that on future visits they can re-explain it. She said, 'we try to make sure they know what it's all about before beginning treatment because it's a serious thing'. Dina was firm about this and told me that she suggests that women who consent to the treatment do so with a full understanding of what is involved in the treatment. One egg seller, Lily, said:

> I'm worried about the risks of the treatment. I'm scared there will be damage to my ovaries. I'm going to be a medical assistant, so I think about the consequences. I asked the doctor about this, but she didn't tell me anything. (Lily, interview, Romania, 10 September 2002)

Hence it is not so clear that the risk of the procedure is fully explained. Further, even if Dina and Iulia do explain all of the risks, it may be the case that some sellers do not appreciate these until much later in the process.

Nevertheless, women do sell their eggs and become enrolled in the clinic's assessments of a 'good girl/donor'. For instance, some women produce more eggs than others and because of this they are repeatedly called back to donate. They are permitted to donate up to six or seven times. From what I observed by looking at the spreadsheets that were running on the computer in the laboratory, some women donate to a number of different recipients – up to five. Some couples receive oocytes from several different sellers. One seller, Valentina, said, 'I will donate again, [pause] if they call me. *I have to be good.* If I make a lot of eggs they will call me' (my emphasis). There is a sense in which in order to be 'good' the seller has to become a 'good', they must acquire value in economic terms. Thus, the sellers become strategically self-objectifying in order to gain acceptance in the clinic (Thompson, 2005). In assessments around the clinic, the quality of the seller, or 'how good' she is, depends on her capacity to 'deliver the goods'.

She must also be considered beautiful. In Israel I was told of the 'beauty' of the sellers here (see Chapter 3), and when I spoke to Dina I asked her whether a seller's appearance was important. 'Yes', she replies, 'it's important how she looks. We have – you will see – we have nice girls. Also we have maybe girls not so beautiful. But, for example, we don't have fat girls.' Beauty, although important, may become less important in the face of the pressure to procure more eggs. However, according to Dina's explanation, this did not hold for 'fat girls', who did not respond well enough to the hormonal injections and were thus excluded from recruitment. She says:

> it's not something [wrong] with us [...] we [perform the treatments] sometimes with these girls. They come, they beg [...] we make with some girls and they don't respond to the treatment. It's not good, if you think like an administrator, you know, with regard to the money; because it's expensive treatment. (Dina, interview, Romania, 8 September 2002)

It would appear that there was a consensus around the clinic that 'fat girls' do not produce sufficient eggs and that, for this reason, they were not 'good sellers'.

Meeting the first sellers

In this round of oocyte extractions there were 20 egg sellers. Over a period of one month the women were injected with ovarian-stimulating hormones. They arrived nightly for three nights at the clinic for this injection, which is administered by a nurse. Six or seven women came to the clinic each night, and each morning six or seven came for extraction.

At around 10:30 p.m.[6] the nurse arrived, tired and dressed in clothing that seemed too big on her. She had spent the day working at another hospital and was now about to administer hormone injections to seven egg sellers. The first seller, Anna, arrived and sat looking a bit nervous. I chatted with her and told her about my research. She appeared to be interested in this and told me about some of her experiences with oocyte donation. As the others arrived they joined in the conversation. Three women could understand and speak English relatively well, so they acted as translators. It seemed as if the young women knew one another, but they told me that they had met only at the clinic.

These conversations I had in the waiting room with the sellers raise many problems. I was asking them about the risks of treatment, and this often made them worried. One woman thanked me because I had 'opened her mind'. This response troubled me. I was not trained to deliver information about the benefits or risks of hormone treatment. I know that it can be problematic and patronizing to intervene to 'improve' the situation of women in other societies.[7]

Even by asking the question about 'risk', I was colluding with very problematic discourses that construct reproduction and women's health as risky. Such discourses are problematic because they require a persistent self-disciplining and surveillance by women, which makes them feel that they are constantly under assessment (Bordo, 1993; Foucault, 1978). Discourses of risk are thus in some sense mechanisms of social control. Yet after observing the clinic's procedures and interviewing the sellers, I felt that some of the sellers were not well informed about the clinic's procedures and its potential dangers. I therefore suggested that they check the Internet for further information.

On the first night of my research in Bucharest, after all of the young women had had their shots, they left. Louise and I chatted with the guard, who was probably in his forties, at the gate for a little while. He appeared worn and tired, but very friendly. He told us to make sure never to take a private-hire taxi and only to use taxis from an established company whose phone number is published in the telephone directory. He seemed pretty adamant about this. He emphasized that two women should be careful travelling alone in Bucharest.

Later that evening, Sasha the embryologist from the Tel Aviv clinic arrived. The following day, he would begin the procedures of 'egg stripping', Intra-Cytoplasmic Sperm Injection (ICSI) (or the process of inserting sperm into oocyte) and IVF. He settled in his room to sleep. The nurse left. Louise and I retired to our room, where we chatted for half the night about what we had witnessed thus far.

Before I went to sleep I wrote in my notebook about the 'dangers' of research in Romania, which I had been warned about by so many people. From what people in Israel and Toronto had told me, it seemed as though I was entering a lawless, violent country. Dina indicated to me that the night guard is a necessary precaution against the 'gypsies' in the neighbourhood who had already stolen purses from the clinic.

An oocyte extraction observed

A black-haired, extremely thin woman was unconscious on the surgical bed. Sasha handed back a test tube through the lab window to Natalia. The seller's hand moved slightly. Her eyes opened a little. When the transducer was removed from her vaginal opening, no chunky red material came out, just a bit of blood. Maria cleansed the woman's genital area with paper towelling from the counter. The seller was removed to the recovery room and another seller with very bright, long, fuchsia-coloured hair entered. Natalia prepared the tools for the next procedure. Iulia and Clara washed their hands and one of them told Natalia to switch off the light. Natalia drank coffee while waiting for the seller to fall asleep. Once the patient was asleep she was disinfected internally with iodine and externally with isotonic saline solution. Much of what was being said eluded me as they were speaking too quickly for me to understand. Clara's phone rang during surgery; she answered it and had a conversation while tending to the woman's arm. She remained talking on the mobile phone for the rest of the operation. Iulia completed the operation and went immediately to ask Sasha how many eggs they got.

In the laboratory

By reading the computer screen, I saw that they had extracted 31 eggs from the third woman. I sat for a bit to observe Sasha. He was collecting eggs and putting them in the incubator. His mobile phone rang and he said, in Hebrew, that he was finishing the fifth *dikkur*.[8] At the Tel Aviv clinic this procedure was often referred to, in softer, more euphemistic language, as '*she'iva*', aspiration/suction, but here in Romania Sasha reverted to the language that I had heard in the state-run hospital in Jerusalem (Chapter 5). Otherwise this procedure was referred to here as 'aspiration'.

Sasha explained that eggs are placed in a medium that is appropriate for conditions outside the incubator and that he then transfers them to a medium appropriate for storage inside the incubator. Eggs can survive for up to 18 to 24 hours before being fertilized – but it is preferable to fertilize them within 12 hours. After separating eggs into the medium, Sasha labelled the dish with the seller's file number and the number of eggs. Sasha explained that he was stripping the eggs. In order to do so, he puts them in enzymes that soften the 'shell of cells' around and then 'peels' the cells away with a very small pipette, the tip of which is the size of an oocyte. The stripping was done with the end of the pipette in the mouth, to create a vacuum by lightly sucking on the pipette. Sasha emptied the contents of the test tubes that Natalia handed him and returned them to her to be reused.

After a while I noticed that all of the chemicals that Sasha was using came from Israel. As well, the chairs that we were sitting on were the same as the ones at the Tel Aviv clinic. The worktables came from a manufacturer called 'Haroshet'. *Haroshet,* meaning mill, industry or manufacture, is an Israeli-based company whose logo approximates the 'male' or 'Mars' Greek symbol (circle and diagonal arrow) in two shades of blue. The red tray holding test tubes had Hebrew writing in thick green marker. Sitting in the lab, listening to Sasha explain things to me in Hebrew and speaking on the telephone in Hebrew, it was as if we were transported to Israel. Or more accurately, it was as if a part of Israel had been transported to Bucharest. Indeed the signs on the doors are all the same as the ones in Tel Aviv. When Sasha left the lab for a moment, I looked through the window behind the worktable and saw people sitting outside a house on chairs in a paved lot – there was a large tree stump serving as a table – one woman washed clothes in an orange bathtub. A dog scratched itself and rested in the shade. The woman hung laundry out on a line. There was discarded stuff around: wooden panels, bathtubs and an armchair. These were the 'gypsies' who were being kept out by the guard downstairs. The differences between these people and their stuff and the people and objects inside this Israeli clinic were striking. The ova, furniture and hormones are able to travel back and forth between here and Israel. The sellers and staff members, except for Sasha and Dr Shmuel, do not travel. The differences between the high-tech inside of the clinic and life outside it were sharp reminders of the mobility and immobility in global repro-migrations.

I went into the lab to see what was going on there. There was another person in the lab, a biologist named Sandra. She did not speak English and our verbal communication was minimal. Sandra was typing 'seller details' into an English version of a spreadsheet program on the computer. She permitted me to watch for a while, and I suspected at one point that she had written in a seller's hair colour incorrectly, 'brown' instead of 'blonde'. I took a short break to eat, convince myself to be less suspicious (the scene with the first seller bleeding heavily had left me slightly anxious), and then return to the lab where Sasha was completing the spreadsheet that catalogues seller blood type, height and hair, eye and skin colour. Sellers' numbers and recipient file numbers were listed on this spreadsheet. I mentioned to Sasha what I believed to be Sandra's earlier inaccuracy with hair colour and he went downstairs to ask Dina. While he was downstairs, I observed that all of the boxes for blood type were blank. I noticed that many boxes in other categories were blank as well. There was a variation in the numbers of slots filled out for other categories. It seemed that the box most faithfully filled out was the one with the numbers of oocytes. Other categories seemed less important as they were not filled out. The first two sellers had produced 18 and 22 eggs, respectively.

Later, I watched Sasha pair up eggs with recipient couples' sperms. Some eggs were divided between up to five couples and some couples received eggs from two different sellers. It was difficult to ask him questions at this point, as his work was extremely careful. However, Louise had chatted with Sasha about this process of deciding who to give which eggs to, when they were drinking coffee together earlier. She told me that when asked about whether he felt that he was creating Life, Sasha said, 'I do. I pick the sperm and put it together with the egg.' When asked what he does to decide how to make a match between egg and sperm, he replied, 'Uhm, a feeling.' With regard to the identity of the seller and the desire of the recipient, Sasha stressed that the recipients were more concerned with skin colour than with the religion of the seller, 'they care whether the child will be white, brown, dark or black. They want to know hair colour, but not religion.' Referring to recipient preferences, he finally added, 'basically they just want to know she's not a Gypsy' (but see Chapters 3 and 4 for an analysis of race and religion in egg donation).

The next day

I began the next day by spending time in the recovery room with the sellers. Prior to surgery they were not permitted to eat, and they talked about food constantly. They were scared and giggled nervously. We chatted. Occasionally I asked pointed questions about donating eggs. One woman told me that she heard about the clinic from a friend. She was in need of money but was hesitant to undergo the procedure. It was not an easy decision for her to come here, 'I was afraid of complications,' she told me. But thankfully the first time that she underwent the aspiration she did not have complications. Nevertheless, she said that she wants children 'very badly', and was worried that somehow her reproductive system would be damaged. She told this to Iulia, and Iulia explained the entire process to her. 'I like this kind of experience,' the seller said. 'I watch Discovery Channel and, and [...] I don't know [...] it is very interesting.'

I knew, of course, to what she was referring. Discovery Channel provides a viewing experience that can be had almost anywhere in the world. The programmes are about topics relating to nature, animals and biographies of famous people. What Discovery Channel offers is the sense of participating in a global knowledge economy. The oocyte seller, it seems, was articulating her desire to participate in something 'larger' than her present life, something more global.

Another woman interjected at one point: 'I was talking to my friend (who is also doing this[9]) and we were thinking, "where are they going, our children [referring to eggs]? How will they look?" I have a sadness in my heart about my future child.' She told me how much she wants children in the future, 'a whole football team'. But, she stressed that she wanted her career first. We then got sidetracked when they began asking me whether I like Romanian 'boys'.[10] However, I returned to the topic of her future children. Her suggestion that the oocyte that she is producing are her future children is interesting. She aspired to having many children of her own, but attached this desire for children to a career, which will help her to become upwardly mobile. Her aspiration to have a career will, in one sense, be facilitated by the aspiration of her oocyte. However, she must detach herself from this particular 'future child',[11] in order to achieve her goal, at some point in the future, of having a child that she can support financially. The hopeful act of oocyte extraction mobilizes the hope for a better life in the future.

Later that evening

In the evening, I talked to sellers in the waiting room while they awaited their final injections. As we talked one woman voiced her fear that the medications might not be 'legitimate', because she had 'asked the pharmacy about Decapeptil [an ovarian-stimulating drug] and they do not sell this drug'. So she had become suspicious. I mentioned that Decapeptil is a standard IVF drug but then suggested that she look on the Internet to find out more about it. Another woman joined in and told us that she worried about the effect that this procedure would have on her ovaries but that the physician was evasive about this. Clearly this contradicted what Dina had told me about explaining everything and clarifying matters for the women.

I asked the sellers why they donate, what led them to donate their oocytes. They told me that they donate 'out of desperation'. They said that they were desperate to get out of constant debt, so that they can buy themselves basic 'necessities' such as clothes, new bedroom furniture, makeup, cigarettes. One woman was behind on her rent for two months and so decided to sell her eggs rather than borrow money. One woman got up and left the conversation. At first I interpreted this as having to do with my questions. But I found out later that she was upset because Natalia would not permit her to self-administer the injection, despite the fact that she was also a medical assistant. A woman mentioned pain in her lower abdomen, and that she believed that she still had symptoms from the previous cycle of donation. Another asked the more experienced women how the procedure is done. Some of the women told me that they have children, others did not and were ambivalent about whether they will ever want them.

Mickey Mouse medicine?

On the clinic walls a series of posters hang. One has been placed in the corridor on the first storey of the clinic and the other on the wall along the staircase that leads up to the second storey (where the surgery room, laboratory and recovery room are located). Sellers cannot avoid seeing these posters, probably for the first time, on their way up to prepare for surgery. As well, the posters are passed by members of the clinic staff on a regular basis as they ascend and descend the two flights of stairs.

These two posters show Mickey Mouse holding flowers. In the first (Figure 2.3), he is standing still and in the second (Figure 2.4), he is running after someone. What kind of offering is Mickey making? Mickey Mouse is carrying flowers in both posters and in both posters he and the flowers are the only items that are shown in the primary colours of red, yellow and green. The background of the posters is a grainy black and white, almost as though it was a photograph that had been 'touched up' or painted. Mickey strikes two different poses. In the first he is coy, while in the second he appears more desperate, eagerly chasing someone outside the frame of the picture. He is making a love offering of a gift of flowers. Some have fallen on the floor and reach out of the frame, like hearts on stems, beneath his oversized feet. The flowers strewn on the ground in the first poster, where Mickey appears calm, may indicate an element of casualness to his offering. Perhaps he has just picked up the three flowers off the floor and offered them, or perhaps they have fallen, but he seems not to notice. In the second image, in which Mickey appears to be eagerly running after someone, the yellow flower in the middle has fallen off its stem.

The black-and-white effect makes the picture of Mickey Mouse in both posters appear as though it is a photo mounted on a white background. This is a rather nostalgic image of the black-and-white era of television and cinema. Yet this nostalgia is undercut, as the fallen flowers exceed the boundaries of the 'frame' of the picture, and the edges of the flowers lie on the plain white background that surrounds the picture. The nostalgia of the 'photo' is undone by the flowers that remind the viewer that the entire poster is a fabrication. On the one hand, viewers are invited to see Mickey Mouse as a 'real' and serious suitor and at the same time, on the other hand, they are reminded that he is only a fabrication and not meant to be taken seriously.

The statements below each image seem to create a relationship between Mickey and someone else, either the viewer or the imaginary character Mickey seems to be pursuing. The relationship is created through the poses that he strikes and his gaze out of the frame of the posters, but also through the writing below the pictures. In large block lettering, the first poster says 'TRUST ME BABY'. This is linked to Mickey's more hesitant pose where he is leaning backward, his shoulder

Figure 2.3 'TRUST ME BABY', Bucharest clinic
Source: The author.

slightly raised and his hand in his pocket. The phrase 'Trust me baby' seems oddly jarring to me in the context of this clinic, particularly when it is accompanied by these images of Mickey Mouse. From one perspective, Mickey Mouse appears to be asking for trust. Yet his hidden hand and the strewn flowers make it appear a haphazard, hedged and untrustworthy effort.

The second poster says 'WAIT BABY WAIT'. Mickey Mouse is depicted as running, with beads of sweat streaming off him, making it seem as though he is running in pursuit of someone. Read together, the two posters illustrate a change of tactic in the 'story' between Mickey Mouse and whomever he is chasing. It would seem when Mickey Mouse's appeal to 'trust me' had failed, that he then had to chase after them.

Figure 2.4 'WAIT BABY WAIT', Bucharest clinic
Source: The author.

In both posters the word 'baby' is tacked on at the end of the phrase. In this instance, 'baby' is a word that is sometimes used as a term of endearment between romantic partners, although it is an oddly infantilizing mode of reference. In the context of this clinic, 'baby' could refer to the sellers, or it could refer (as sellers often did themselves) to the eggs as if they were babies. (This collapsing of babies and eggs into one another is also examined in Chapter 3, where I discuss the phrase, 'embryos are our *baby*'). There, I suggested that 'baby' could refer to the clinic's project of making profits through promoting a hope in the technologies of IVF in order to divert attention away from the relatively low success rates of the technology. In this respect, 'baby' could refer to the aspirations of the clinic, which I elaborate on below.

The wording and the offering of flowers makes the posters quite traditional and gendered images. Mickey is a masculine figure, perhaps representing the doctor or embryologist. And obviously in this context he is communicating with the sellers. The gendered aspect of the image coupled with the sexual or romantic overtones of the words below may set up a slightly unsettling relationship between the intended viewer (the sellers) and the presumed referent for Mickey (the heads of the clinic, Dr Shmuel and Sasha). In some sense this storyline, which changes from a casual to a more active one, foreshadows some of the features of the operation of the clinic that I observed. While the doctor relies on the sellers, and may take for granted that they will be an available resource for his enterprise, Dr Shmuel was sometimes required to exert a certain amount of energy in order to keep these sellers.

Mickey Mouse is a comical character that often represents the multi-billion-dollar corporation, Disney. These images of Mickey Mouse are intended to evoke associations with humour, and to calm the sellers on their way up to surgery. Yet the phrase 'Mickey Mouse' can also imply 'ineptitude'. Given this connotation, the posters are an odd choice. While Mickey Mouse 'offers' the Disney fantasy of love, happiness and a happy ending (Project on Disney, 1995), the phrase 'Mickey Mouse' may actually evoke unintended negative responses.

Furthermore, as Mickey Mouse is also associated with corporate capitalism, the images suggest the clinic's affiliation with and location in consumer capitalism. What Disney offers in the figure of Mickey Mouse (love, freedom and a happy ending) is the unattainable dream of many Romanians. A few days after my research at the clinic, we stayed at a small hotel in the centre of Bucharest. The man who runs the hotel, Dan, drove me to the airport on my last day in Romania. On the way he talked extensively about the kinds of things that he – and according to him, most young Romanians – dreams about, including: buying bedroom sets for their homes and other kinds of furniture. Dan emphasized to me that it is insufficient to have food to eat and a car to drive. What Romanians crave, according to Dan, are the accoutrements that they believe come with a Westernized democracy – the goods of contemporary consumerist Western societies. Many of the sellers talked about purchasing goods for their homes, such as new furniture. Others wanted to buy clothes or to redecorate their homes. As the sellers' narratives suggest, the need for money and *stuff* indicates their desire, not only for the 'good life', but for a Westernized good life for which Mickey Mouse stands. This is not to say that they wholeheartedly embrace neoliberalism, but that participation in what is seen as a wider global 'culture' and having a 'better life' are central desires to the people whom I encountered. The posters of Mickey Mouse locate the practices and stories that I observed and heard at this clinic within transnational movements of people and objects (Appadurai, 1991). These repro-migrations may allow for the 'flow' of some objects, body parts and people while they block the movement of others.

Romanian oocyte sellers are theorists of reproduction

Laura

Laura is 22 years old and works at a firm where she has a very busy schedule from 8 a.m. to 5 p.m. At the time of our interview in September 2002, she was a two-time seller. Her previous oocyte extraction occurred four months prior to our meeting. She used the money from the first cycle to buy a television and vacuum cleaner. This time she will redecorate her kitchen. She told me that her salary just covers the cost of her food and other necessities. Laura explains to me that it was not an easy decision for her to make, to sell her eggs. She thought about it for a month, and consulted with her boyfriend about it. He told her that it was her body and that she should do whatever she wants with it. 'I don't want to have kids now,' she tells me,

> Because it's hard to live. Raising kids is complicated and you have to give them education. He must grow in a perfect place. You have to offer him everything. And I cannot do this now. And I think if somebody needs my eggs to have a baby and raise it properly I think it's good that I'm doing it. If I can help other people, why not? (Laura, interview, Romania, 11 September 2002)

Some might say she's suffering from a form of false consciousness. But she is highly intelligent and has made a financial calculation regarding the selling of her labour power/body parts: like the other oocyte sellers, she is paid US$200 for undergoing one cycle of 'aspiration'. After the surgery she will receive an envelope with the US$200, in cash. This money amounts to twice her monthly salary. Despite the fact that she said that she would not do this again due to the pain that it caused her the first time, she commented that, 'It's addictive, like a drug', and 'it's a constant need for money. It's something you need, the money. For money you can do almost anything.' Dina the clinic manager (and her friend) told her that in the USA they pay a lot more, 'but how do I get there? It's all about money.' Her calculus includes a sense of her own immobility.

The 'right' to do with her body what she wants – a 'pro-choice' discourse – is put to use in this context in order to justify selling oocytes. She recoups her alienated labour as a 'right'. Taken within the context of the entry of Romania into the European Union and the attempts of the Romanian state to become a European hub of business and tourism, she is performing modern Western citizenship here. How can 'dignity' and 'human rights' be invoked when selling a body part is turned into a 'right'? According to the Kantian notion of 'dignity' (upon which Western notions of dignity are based), oocyte sellers are excluded from having dignity because they have a market value (Khanna, 2007: 263–4).

Elena

Elena was a 24-year-old woman who worked as a manager selling used books in the underground markets of the metro. She was a part-time student of construction/building, and she wanted it to be known that she studied this in French. She was a second-time seller. Elena heard about this clinic five months prior to our meeting, from her friend who delivers bed sheets to the clinic from a linen company that is based across the road. Despite the fact that she reported that the treatment felt easy to her, her employment made it difficult to schedule sessions at the clinic.

She told me that her family did not know that she was doing this because they live far away and, 'they are old and don't understand' and she was not close to them. Besides which, 'my mother is very, very curious. If she knew, she would come here to see [...].' Elena has told her friends, however, and they were going to the clinic to sell their oocytes as well.

Elena informed me that she was aware that there are certain risks in donating eggs. She added that she did not feel that there are big risks involved in the treatment. This made it much easier for her to undergo. Furthermore, she had distanced herself from her relationship to her oocyte, saying, 'it's not *mine* mine'. That is, she did not consider the child that could potentially result from her donation as her own. She denied any kinship relationship with her oocyte. Hence, she had decided that she did not want to know about what she nonetheless referred to as, 'my baby'. She adds that she hopes that it will be okay.

She said that she did not want to sell them a second time. However, Elena's need for money made her do it. Perhaps Elena has detached herself from her oocyte because of this. She said:

> I feel shame to win the money in this way. I believe it's not a help for me, because I don't really work for this money. I prefer to work for this money, but in this case it came and it was easy. It's like a weakness. (Elena, interview, Romania, 11 September 2002)

In Elena's mind this is not a form of labour, but a sale of a body part. She associated not working for money as negative. She told me that she would not donate again but that she would use the US$200 that she received to pay her rent, which is three months late. She planned to buy clothes, to go drinking with friends and to give her mother a portion. Doing some arithmetic in front of me, Elena determined that she was paid 32 cents an hour at her job. Despite this low wage, she counted herself lucky because she determined her own work schedule and could thus accommodate her studies. Having the freedom to decide, whether to sell or not to sell, what to spend the money on and whom to tell about it seemed to be the crux of many of the sellers' reasoning about donating. She is aware of the deeply exploitative relationship that she is involved in, and recognizes the lack of real meaningful choice for her in this situation.

Vanessa

Vanessa was a 19-year-old first-time seller. She was married, had a two-year-old son and lived in Bucharest. She worked in a factory that made Barbie accessories. Her job was to assemble the boxes in which dolls and accessories were packaged. When I first met her in the waiting room, she had a large pink Barbie box with some sort of Barbie accessories inside. The women in the waiting room (including myself) were all talking about it and passing the box around to examine it. Vanessa analyzed her situation through questions of health risks, future fertility and, like previous women, a comparative analysis of the value of her promissory capital (Thompson, 2005). She initially worried about how the treatment would affect her health, but her friend advised her that she could trust the clinic, that it was 'Okay'. Vanessa was worried that she would not be able to have another child because of this procedure. But her friend assured her that, two months after the treatment, she would be able to have a baby. She waited to see the effects on her friend. After the treatment, she saw that indeed her friend felt well. Based on this, she decided to do it as well. Thus far, she had not had any side effects from the medicine, but, she added, the procedure made her stomach hurt a little. I asked her why she had decided to sell her eggs. She told me that her decision was 'for the money'. She explained that the money that she receives from doing this procedure is quite good. It amounts to four or five months' salary. She quickly added that part of her decision to donate had to do with the fact that her uncle and aunt cannot have children and are very unhappy. She said that she thinks of them when she sells her eggs. She added that, unfortunately the eggs are going to people who can afford to buy them, rather than to poor people. But she believed that it was better for the oocyte to go to people outside

Romania, 'because here it's poor' and the children that result from the oocyte donation would have a better life if they left Romania.

The narratives above are 'reproductive moments', where a desire or need for money is what gets narrated and therefore is what is reproduced.[12]

Although they play a vital role in helping some Western oocyte donation clinics to function, the women egg sellers are distanced from the people who run the clinic and the oocyte buyers/recipients both socially and economically. Their own aspirations to improve their lives are attained through a detachment from their body parts. In this sense, some of their material aspirations are enabled by this detachment. The aspirations of Western scientists, of IVF patients and of egg sellers may be viewed as being related to each other through an oppositional or 'torqued' relation.[13] There is an inherent inequality in the relation, to which it is not reducible. It is a temporally limited global kinship, produced by/in/through the desire for capital, freedom, migration and children.[14]

I suggest that the multiple desires of sellers and recipients to give and receive eggs in Romania and Israel produces a kind of relationality that is temporary but noteworthy. 'Aspirations' (extraction of oocyte) may well be 'reproductive moments' (Strathern, 1995), in the case of Romanian oocyte-sellers. Their 'intent' and the intent of recipients may be what replaces biological ties and creates a kind of 'torque' between them. The sellers' decision to donate may indeed be linked to the sense of having 'no choice' (Franklin, 1997) but to donate, because of economic constraints and aspirations. The oocyte selling stories indicate these momentary 'nodes of desire' (Nahman, 2008), where one may attempt to gain a sense of dignity within global capitalism by doing precisely what perpetuates the system, producing, buying and selling.

It is important to emphasize that these repro-migrations do not completely and permanently immobilize sellers. The sellers have other desires, which will create different kinds of relations. These will have different effects on their mobility in an economic and global sense. These may include, for example, the social mobilities that transpire as sellers gain educational qualifications (as some of the women in my study were doing) or access to migration that is so prevalent throughout Eastern Europe (Wallace, 1999: 199–200). That is, the problem here is not the process of reproduction and reproductive technologies themselves, but the system that enables the inequalities that pervade and persist.

Discussion

A feminist critique of bioethical debates around human egg markets has been that the 'lady vanishes' (Dickenson, 2006), that is, women's contribution of the egg, material stuff that leads to babies or stem cells, was not central to bioethical discussions. Yet bioethicists are rarely afforded access to firsthand ethnographic observation and interviewing. Meaning that often even within these discussions, the woman who sells the eggs, her own perspective and specific historical and geopolitical positioning is absent. The other social actors also 'vanish': doctors, nurses, embryologists, lab technicians, donor recruiter; and technologies, spaces and objects: aspirator machines, ultrasound probes, surgical beds, waiting rooms and all of the encounters that can occur in those spaces. This raises the question of what an analysis of global ethics of egg donation under global capitalism could look like when given ethnographic access. A feminist science studies,[15] which looks at the 'enactments' of practices of oocyte extraction (Mol, 2002; Barad, 2007), or an anthropological one that privileges women's wor(l)ds affords an opening up of spaces of deeper understanding. Hence the central concepts of analysis for global ethicists of human markets in tissue – exploitation, choice, trafficking – become elements in an ethnographic unfolding of stories. In this chapter there are echoes of trafficking and anti-trafficking protocols, there are moments when things or people travel while others do not. There are clear instances where anti-trafficking laws, and specifically the 'force' aspect of exploitation in trafficking, are clearly being breached in this clinic (Nahman, 2011). Questions of 'trafficking' and 'reproductive labour' in global egg markets, which are often answered with policy, or secondary data analysis can have different answers when expanded with a few more layers of analysis: first, the voices of the oocyte-sellers themselves; second, the cyborg-anthropology configuration of non-human actors and fictions (Downey and Dumit, 1995; Haraway, 1991); and third, an analysis of the researcher's own emotions, observational capacities and perspectives in the production of an egg-donation story. When seen ethnographically, the view of 'exploitation' and 'choice' can be unpacked through the women's own interpretations of these things.

Extracting oocytes out of women's bodies is mining bodies for the creation of monetary value, through the development of new techniques, generating more research papers for scientists and developing new pharmaceuticals associated with reprotech. This is made salient when we view the extremely low success rates of IVF with oocyte donation (Nahman, 2007; Throsby, 2004). That is, if we take into account the fact that very few babies result from IVF with oocyte donation, the reasons that so much energy is put into the practice (setting up transnational clinics, sending people, gametes and medical equipment around the world) become much more about something *other* than the reproduction of humans.

Yet it does not follow that the women selling their oocytes are simply exploited, either. They are savvy participants (even entrepreneurs) in this capitalist economy, where desire operates as a force for linking differently positioned women. Eastern European egg sellers are not *passive objects* at the mercy of global capitalism, bioenterprise and the desires of egg recipients. The oocyte sellers are actively engaging in selling eggs. However, whether they are active or passive somehow does not capture the nature of the relationships in this system. The no choice of 'choice' is plainly evident in egg donation as it is in IVF (Franklin, 1997).

Consent is a slippery concept, where patients do not always know what they are consenting to, to what future use their body parts will be put and in this case, where their oocytes would be taken (Corrigan, 2003). Some sellers with whom I spoke immediately before their extraction was performed, expressed persisting concerns about the process of ovum extraction. The women I spoke to, whether highly educated or not, were aware that the process that they were entering into was complicated and that it put them in a somewhat compromised position. To speak here of 'agency' (that neoliberal fantasy of individualized, autonomous action) would lose out all of the historical, collective and emotional dimensions of this process. Charis Thompson writes about women who engage in an 'ontological choreography' through which they place their own subjectivity aside briefly and become 'objectified' under the surgical knife in order to become new kinds of agentic persons: mothers (Thompson, 2005). The 20 Romanian oocyte sellers whom I interviewed are in some ways very different from the IVF patients whom Thompson presented. Oocyte sellers are differently positioned to one another, in terms of their relationship to the state, power, the global economy and oocyte recipients. The women whom I interviewed are solely sellers, so that the benefit that they will gain is monetary. Second, these sellers, in post-communist Romania, are positioned differently to USA oocyte sellers in a global economy. There is a unique kind of collective ontology going on in this clinic, which is also about sellers changing the role of gender in their society. As a process of inter-cumulating, selling eggs is also a way of changing their national and global standing.

As a way of explaining this further I turn to my chosen name for the women that I interviewed: sellers. To call the women that I interviewed 'donors' would be a misnomer. They are explicitly there to sell their oocytes for a specified sum of money. The way that desire operates for them in this economy is about getting a supplement to their income. She gains something as a woman and as a Romanian by being able to participate in this economy, where some people want to reproduce and she wants to supplement her income. We may then need another language with which to articulate the situation where oocyte sellers are something other than brutalized victims who need a helping hand to be made to feel dignity, which is what is implied in some of the feminist research on the topic. Perhaps the historical and geographical specificity of this example is useful for careful assessment of the moral terrains of these practices.

Watching eggs being extracted from women's bodies made me realize that the 'victim' not only has a voice but that she herself may be a theorist of reproduction (the anthropologist Martin (1995) looked at HIV patients as theorists). She not only navigates a moral dilemma of how or whether technology should be used on her body (Rapp, 1999), but she also decides to be someone who sells a partible part of her body. I am not arguing that her 'autonomous' 'choice' exempts her from being exploited or trafficked (see Nahman, 2011). Indeed, neoliberal choice is a fallacy (McNeil et al., 1990; Franklin, 1997). Rather, her sense of her own autonomy is a vital part of the story of how globalization and capitalism work on people. Sometimes not making a baby, but selling an egg can be seen as emancipatory. Capitalism takes advantage of such niche markets. Israeli oocyte extraction in Romania involves complex sets of practices, motivations and aspirations.

Yet, oocyte donation in Romania is also part of what makes Israeliness.[16] The nation is constructed from inside the borders of the state, through practices that keep some things and people out and keep others in. However, it is also made through transnational movements and practices. This mobility is a structuring aspect of contemporary Israeliness. Romanian extractions in this sense *produce* Israeli extraction. Here we see the ways in which the aspirations of the seller and of the practice of aspiration intersect. In this instance the seller's aspirations were to donate eggs, receive money and still be healthy. The clinic's aspiration was to extract the eggs. These two aspirations are materially and semiotically intertwined in the 'Blood in the Aspirator' story through the blockage and flow of blood.

3
EmbryoMethod

A brand and a 'method'

In the Universal Fertility Centers (UFC) company logo, a faintly purple cloud of an embryo floats inside a purple sphere, encircled by two white, elliptical lines. It signals the global aspirations of the company, its association with high-tech futuristic thinking and the centrality of embryos to its practice. This logo is found on all company letterhead and on office doors in its five Israeli branches as well as in the Romanian ova-extraction clinic. It is an attempt to 'brand' the company, to create an immediately recognizable mark that will be associated with the company, such as the Nike 'swish' or McDonald's golden arches. Whilst at the time of my research this was primarily an Israeli company, it now has strong links with the USA and Europe, with clinics in other parts of Eastern Europe as well as Israel and Romania. Its global aspirations are being realized.

Transnationalism makes nationalism

The examples in this chapter are drawn from research in a private IVF clinic, south of Tel Aviv. This clinic is part of a network of six clinics, known as the UFC. It is located in an industrial zone of one of the Tel Aviv suburbs and is housed in the building of a large chain of privately owned hospitals that has branches around the country. The network of clinics was spearheaded and linked together by two people, Dr Shmuel (the head physician) and Orit (the head manager). Five of the clinics are located in various parts of Israel, in the areas of Jerusalem, Tel Aviv, south, south central and northern Israel. The sixth clinic is in Bucharest, Romania (see Chapter 2). Each of the Israeli clinics is run by a manager, who organizes patient visits, meets with prospective patients and generally ensures the smooth operation of the clinic. In addition to this, each of the five clinics has a nurse, and an attending physician who visits once or twice a week. These physicians are usually linked to other institutions (generally state hospitals) in which they practice. The work that they take on at this clinic is additional to their other medical obligations. The network of ova donation clinics was imagined by Dr Shmuel as an 'all-in-one solution', to meet the desire of couples for children and has been running since 1997.

While specializing in egg donation, the clinic also provides surrogacy arrangements and IVF, and Dr Shmuel also considered adding a transnational adoption service. The satellite clinics around the country are centrally run via telemedicine working on an 'expert system', which Dr Shmuel co-designed and was in the process of patenting. While his satellite clinics are for intake and prescription purposes only, the central clinic is where the clinical procedures are performed (uterine investigations, embryo transfer and extraction of eggs in the cases of surrogacy). On a monthly basis, at the oocyte extraction clinic in Bucharest, approximately twenty ova donors in Romania are hormonally induced to produce a large number of eggs (usually between 15 and 30). An embryologist from the Israeli clinic travels with the head physician to Romania, carrying with them a canister of cryopreserved sperm from commissioning couples. In Romania, the sperm is thawed, ova are fertilized and incubated. When embryos form, they are frozen and taken to Israel where they are transferred into Israeli women's bodies.

On leaving the clinic one day I stumbled upon this graffiti sprayed just outside the clinic, in the parking lot (Figure 3.1). The writing in black says, *'Eyn 'Aravim, eyn piggu'im'*, [If there are] no Arabs, [there are] no [bomb] attacks'. The writing in blue and yellow says, *'Mavet la-'Aravim'*, 'Death to the Arabs'. The violence of the ethnic nationalism quite literally infused the clinic inside and out. Perhaps counter-intuitively, globalization, and the Israeli state's participation in it, does not override or weaken the state and nation (Aretxaga, 2003; Comaroff and Comaroff, 2000; Trouillot, 2001). The difficulty in assessing this, however, is that states and nations are also collective illusions, 'the reification of an idea that masks real power relations under the guise of public interest (Aretxaga, 2003: 401). For instance, the idea that the State of Israel is the outcome of a need precipitated by the Shoah is a collective illusion that masks the historical specificity of the creation of the State of Israel. But to call the state 'illusory' is insufficient, it is powerful and material, as the graffiti on the wall outside the clinic shows. Death to Arabs is not just a collective illusion but also something enacted daily that makes the state what it is. The state is materialized, enacted (Mol, 2002; Nahman, 2006; Ferguson, 2002). Also, saying that the State of Israel materialized because of a need precipitated by the Holocaust and global anti-Semitism is an act that makes the State of Israel a place created out of violence that (in the minds of most of the Israeli collectivity) sanctifies other kinds of violence (Butler, 2004). In order for states to exist there need to be both imaginings and concretenesses, which are enfolded (or entangled; Barad, 2007) within one another. The drawing of the doctor's son of the 'terrorist' inside the clinic (Chapter 1) and the graffiti sprayed by people just outside the clinic are central to the banality and pervasiveness of the violence of the state in everyday life.

As the clinic's most recognizable sign, however, this image evokes 'death'. How we know to recognize that embryos comes from the embryological research that photographed so-called 'dead' embryos (Morgan, 2009). Taken together with the image of the terrorist in the clinic (Chapter 1) and the graffiti outside, death haunted this clinic. ''*Am Israeli Hai*', the hope for the nation to live has unwittingly been coupled here with dead embryos as companions.

In clinic brochures where the logo is printed on each page, another constant is found: the repeated phrase 'the embryo method'. The 'embryo method' has different valences at this clinic that were hard to pin

Figure 3.1 Graffiti outside an IVF clinic in Israel
Source: The author.

down, but they seemed to refer to an 'expert solution', a regulated set of processes and tools, an ability to select desirability globally:

> Based on the *unique* treatment approach of 'the embryo method', which guides the practices of the UFC, there's a *solution* for those hurdles that are well known to any couple who has experienced fertility treatments. The centre functions in a comfortable, discreet, family atmosphere, with much attention to explanation and personal attention [...]. The treatment will be matched to your needs in a manner that will not affect your daily routine and your work. [...] In case it is necessary, you will be referred to the clinic's psychologist to receive emotional aid, support and professional advice. (Excerpt from UFC clinic brochure)

The clinic is making a promise to the nation, they will give people babies 'and fast!' They will make it easy and be the family to patients that the state is not. The 'embryo method' in this representation of it is about logos, brands, wrapped up in notions of 'family', 'embryos', unique solutions. The patient is cared for like an embryo. It invites couples to put their trust in the staff and their technologically facilitated expertise. These treatments are said to be tailored to an individual's needs and routines, whereas it is contended that IVF as practised in other places disrupts work–life and does not meet the needs of the *individual*. The clinic seems to be presenting itself in vast contrast to the militarized medicine image (Ivry, 1999; Weiss, 2002) of the state-run hospitals, where an individual's needs are the needs of the collective, and the embryo is a national one (see Roberts, 2007), foregrounding its rigorous private service that can only be derived from personal treatment and an 'expert system' such as the 'embryo method'. This paternalism indicates the clinic's own brand of doing it for the nation, enlisting the national and individual hopes of prospective patient-customers (this chapter and Chapter 4).

When after reading this brochure I asked the secretaries and nurses to explain what the 'embryo method' is, they did not know. Orit, the head manager, told me that the term 'embryo method' referred to the 'entire system' that is focused on the embryo. She did not say that it was focused on the baby as one might expect from someone running an IVF clinic. I interviewed an embryologist in the laboratory and asked her to describe this 'system' to me.

From the embryologist's account, the 'embryo method' refers to the procedural, bureaucratic aspects of egg donation. Documenting, labelling and transferring. These are state-making practices as much as they are baby-making ones. I arrived at the clinic one morning at 9 a.m. to find that the procedures were only beginning at noon because the anaesthetist had not yet arrived and the doctor would not be able to come in until later that the day. With a few hours to spare, I went into the laboratory to check my emails and chat with Tal, the embryologist. I decided to ask her about herself and about the procedures involved in sending sperm to Romania and in bringing back embryos. I hoped that she would be able to tell me what the 'embryo method' actually was.

Tal told me that she had been a medical student for five years in Russia. She had one year left of her studies but had to cut them short because her husband and family decided that they were to 'make *aliyah*'.[1] She says that she really did not want to leave Russia initially but that she is now very happy that she did. When she came to Israel she had problems in qualifying to complete her medical studies (her studies in Russia were not recognized in Israeli institutions) and so she did a Master's degree in human embryology. She has worked at UFC for two years.

After telling me about herself, Tal begins to explain the procedures of ova donation. She tells me that first the sperm is frozen. Since the couple does not go to Romania, the husband's sperm[2] must be frozen and taken to Romania by one of the embryologists. It takes 40 minutes to freeze sperm and it can be kept frozen for five years and still be 'viable'. When Dr Shmuel and Sasha, the chief embryologist, go to Bucharest they take the sperm with them in a canister. Each vial bears the name and national identity number of the man whose sperm is contained inside the canister, as well as the date of freezing of the sperm or embryo. That the sperm and embryo vials all bear the *mispar zehut*, the ID number (on the national identification cards that Israeli citizens have to carry) of the man or woman (the recipient couple), codes them as belonging in some way to the state, despite being in a private clinic. Unfortunately, I did not think to ask at the time what happens in the case of patients who are not Israeli citizens, have Palestinian identity cards or lack one altogether.

Migrant labourers or people coming from other countries for treatment do not possess identity cards. Furthermore, Palestinians from the Occupied Territories have a different kind of identity card, which clearly marks them as 'other'. I did not have an identity card either and was constantly asked for my '*te'uda*' (pl. *te'udot*; *te'udat zehut* or ID card) everywhere I went. Not actually having one, but being fluent in Hebrew (with the dominant accent) I always confused guards, so I would produce my passport in such instances. Since this is an Israeli passport it was always accepted. It seemed as though the request (at entrances to shopping malls and bus stations) of '*Te'uda be-vakkasha*' (translating roughly as 'ID please'), is a performative statement. Its repetition is what gives it force and what creates the person as a kind of citizen. A *te'uda* is a materialized aspect of the state and is such an integral part of a person in Israel, because it has to be shown or its number provided many times throughout a typical day. Here we see concretely how the individual is a synecdoche of the nation, and the *te'uda* is a synecdoche of the state. You can be asked for a *te'uda* when entering a shopping mall, attending a doctor's office, applying for cable television service or topping up your credit on a mobile phone. It could be argued that the *te'uda* is simply one more aspect of the person in Israel, just like a part of their body. Some individuals, Palestinians from the Occupied Territories or foreign workers, for example, have *te'udot* that mark them as the wrong kind of Israeli. The ID numbers on the test tubes not only help organize the vials and maintain correct handling of the biological material (if this was the only reason for it, a name or file number would probably suffice), but they also mark the substance contained inside as belonging to (both in the sense of property and being a 'part of') the person bearing the number. By putting the recipients' Israeli state ID numbers on the vials, as was done in this clinic, the non-Israeli aspect of the donated material is effaced. The Romanian egg (and its preferred 'race') becomes a part of the Israeli national body, whilst the body of the donor ((her undesired citizenship and religion) is extracted. In this step of IVF, the state, clinic and laboratory are *incorporating* the egg, in the sense of making corporal or bodily the substance in the vials.

Each vial of sperm also corresponds to a sheet with details such as the sperm count, date of freezing, number of portions (test tubes), date taken to Romania and details about where the vial is located in the canister. The mini canister that is taken to Romania is filled with liquid nitrogen and two people insert the vials into the canister in order to be certain that it is done accurately. This occasionally involves three people when there is a *mashgiha* – an Orthodox Jewish woman trained to oversee IVF procedures so as to ensure that they comply with Jewish law concerning reproduction.

Then the sperm is taken to Romania by Dr Shmuel and Sasha (Chapter 2 describes what happens to them from there). The second stage, Tal informs me, is the arrival in Romania, where there are donors who have 'ovum pickup'. There is a file stored in Romania for each donor. All of the donor's details including age, eye colour, skin colour and hair colour are noted in this file (except that as Chapter 2 showed, these details of the donor were not input by the biologist). These are the racial markers that are affixed to eggs and donors (discussed below and in Chapter 4). After the eggs are extracted,[3] donors and recipients are matched. Dr Shmuel and Sasha do this matching according to the patient request sheet that they have brought with them. Then they simply start to divide up the eggs among intended recipients. Egg recipients must wait three months until a 'matching' donor is found.

The sperm is then thawed, and a decision has to be made for each recipient as to whether to conduct an ICSI (Intra-Cytoplasmic Sperm Injection, inserting the sperm into the ovum using a 'micromanipulator') or just do a 'regular' IVF (as Tal said, 'just drop the sperm into the petri dish and then the sperm does his work'). Sixteen to eighteen hours after ICSI or IVF, a fertilization test is conducted, at which point a successful fertilization will show a zygote, or a 2pn (2pn is 'two pronucleii', this indicates that the two cells have combined but not completely, since the new cell still contains two nuclei). If at this stage it is at 1pn, 3pn or 4pn, then the fertilization is not 'normal' and has to be thrown away.

She explains that each ova recipient receives a form noting her donor's 'details', how many eggs she has received and the file number of the donor so that she can check on the details herself. On each recipient's form, next to the number of eggs, they then write the number of 'good' fertilizations. When the zygote is at 2pn, it is cryopreserved and placed in a test tube with the rest of the 'embryos'.[4] The test tubes are labelled with the freezing date and the number of embryos is recorded. At this stage they are reinserted into the canister.[5] When the embryos are returned to the clinic they are transferred to a specific location in a larger container of embryos (this, again, is checked with a second or third person).[6] Once the embryos are in Israel, the recipient woman is 'prepared' for embryo transfer. When the recipient woman is 'ready', the 'embryos' are thawed.

In this instance there are two possibilities, Tal explains. Either the embryos (or rather 2pns) will survive the thawing process or they will not. This depends on many factors such as the methods of freezing, thawing, quality of the sperm and quality of the oocyte. The surviving embryos will be grown in a constant environment in an incubator containing CO_2 (carbon dioxide 5.5%) at a temperature of 37°C. They grow the embryos for three days, and then for two more days (a total of 72 hours) and check their progress – whether they have divided and their quality. The embryologists, along with the physician, then decide how many embryos to 'transfer' and proceed with the 'transfer'. Once the embryos are transferred, the recipient is given hormones to 'support' the intended pregnancy. Two weeks later they conduct the 'beta test' and if it is positive the ova recipient is told that she is pregnant. I observed the embryologists carrying out these procedures in both Israel and Romania. The processes of getting sperm, freezing sperm, thawing sperm, getting eggs, combining them with sperm, freezing the fertilized ova and returning them to Israel are complicated. This process involves classifying and categorizing people and body parts.

In the first instance the 'embryo method' (the term used in the company's brochure) was a brand. It was simplified and codified into a product that patients could buy. The idea of an 'embryo method' shifts attention away from the likelihood that these procedures will fail. If the patient focuses on the promise of efficiency then the extremely low success rates of IVF with ova donation becomes obscured. At the same time the 'embryo method' was a way of private medicine helping recipients achieve a national goal; making patients into embryos cared for by the clinic. The second iteration of the embryo method, the embryologist's description, is a racialized anatamopolitics that parses out the recipient and donor their parts. It labels, codifies and incorporates them into the national body in ways that not only echo state practices but *are* state practices.[7] The third version of the embryo method derives from the head physician's own description of his clinic and seems to represent the ways that global capitalism infuses Israeli egg-donation, facilitating a racialized, national and private project that like the Israeli government invokes Jewish idioms to support the nation. This version is about the clinic trying to distinguish itself from the cold, distant way in which state hospitals run things, but in so doing also reaffirms the state's goals.

> There were several more suicide bombings this week – many soldiers got called to reserves [reserve duty]. They've taken over Ramallah; have fought with Palestinians in Bethlehem. Palestinian families are now sitting at home under curfew with the corpses of their loved ones beside them. And we sit here in Israeli cities and go on with life as usual. The IDF shot at journalists and on this side of things I haven't been able to get hold of more information on what is happening in the Occupied Palestinian Territories. (Fieldnotes, 5 April 2002)

Dr Shmuel contrasted his clinic to state-run clinics arguing that at state-run hospitals women must have various tests conducted, both in the hospital and at their own sick fund clinic, and must be very active in getting test results and pursuing treatment.[8] He claims that his clinic makes things easier for ova donation patients by 'provid[ing] all the treatments under one roof'. Dr Shmuel explains that 'she does not need to *chase after* forms and results, because the UFC *does it all* at their "satellite" clinics'.

The account that Dr Shmuel provided me of his clinic is one of ease and efficiency. He positioned his patients as consumers with a *right* to *convenient* treatment. This is a model of privatized medicine that yet again complicates our understandings of the supposedly benevolent relationship between the Israeli state, nation and the individual body.[9] Ease, speed and affordability are modes of engendering hope in the success of transnational egg-donation. The specificity of this is crucial. Dr Shmuel mixed 'religious' references with allusions to 'national' goals of the state.[10] His account below *combines* ideas about class and the economics of IVF and ova donation with ideas about the selection of donors, their 'availability' and cultural ideas about phenotype. *His* economy of ova donation is not solely consumerist. Rather, the claims that he makes about his motivation in performing ova donation combine a consumerist model with references to aspects of Zionism such as: struggle, selection and the right of return (Nitzan and Bichler, 2002). It is private and national. Or there isn't the same split between the public and the private in Israel as there is in other Western countries. Here the private is also done for the public good (at the same time as it is for personal profit). 'Good citizens' can afford to pay for treatment. He is imagining egg donation through the nation. As we drove from one of his clinics in the south back to his central clinic, Dr Shmuel discussed his clinic with me in a way that laid out some of these specificities:

> Now the main factors for this are the availability of willing donors plus the wide selection. Instead of being limited to one donor every now and then, I can recruit donors, sort them and decide 'this one yes, this one no' based on things which are very significant, like genetic diseases in the family, age (since we know that the chance for success depends on age). The younger the egg, the higher the chances of achieving a pregnancy with it. Whereas in other clinics people undergo ten, twenty rounds, I give them one or maybe two rounds and achieve a pregnancy. Plus we have the ability to match [appearance-wise] the donor and the recipient. And we see the results. And we see that the children who are born look a lot like the people who bear them. (Dr Shmuel, interview, Israel, 11 July 2002)[11]

The idea that Dr Shmuel can recruit 'willing donors' – a 'wide selection', whom he 'sorts out by traits such as hair colour, eye colour and skin tone' – demonstrates how having the global market facilitates a deeply racialized practice, similar to that in other places such as the USA. Yet the coupling of this model with references to the religious and national specificities of Israel makes his description of his clinic distinctive. On another occasion, while driving with Dr Shmuel from one of the other branch clinics, we spoke again about his business. He told me that initially, he did not consider his business to be an 'egg-donation enterprise'. Yet the clinic's glossy brochure materializes this consumerist model.

In his explanation, Dr Shmuel alludes to the effectiveness of his 'embryo method'. He does so through the language of expertise and through reference to 'state-of-the-art' equipment. Yet both modes of explanation shift the focus away from the *effectiveness* of treatment, towards *intended* effectiveness. This is done by an emphasis on the tools for achieving pregnancy rather than on the actual results of 'take-home babies'. The focus on the techniques and tools, rather than the results indicates that the important aspect of the process for Dr Shmuel at least, is his *intention* to assist. The focus is on progress through scientific assistance (cf. Franklin, 1997), which is obvious from his use of phrases such as 'state of the art'. This is a 'positive assessment of natural science in the service of the natural family' (ibid.: 93), and in this case, the national family as well.

The contrast between the tendency of IVF to fail (Throsby, 2004) and Shmuel's description are compelling, indicating as they do some ways in which private medicine mobilizes hope for a baby as both a personal and national endeavour. His description and historical account of the clinic is an ethnographic site that makes this explicit. He echoes and extends some of the promises made in the brochure. The company brand of the 'embryo method' encourages putting hope both in the embryo itself and in the highly technical practices of IVF. The processes of getting sperm, freezing sperm, travel, thawing sperm, getting eggs, combining them with sperm, freezing them and returning them to Israel are full of moments where things can go wrong, require additional tests, overseeing by practitioners and recipients' and donors' obedient participation in injection and hormone intake. This version of the 'embryo method' then is a use of private capital to coordinate things, people and processes to make things *work* together. The company brochure taps into the growing consciousness in Israel that the welfare state is not working. It uses the language of simplicity and ease to distinguish itself from the nationalized health service providers where things might go wrong (because the state has failed to invest sufficiently). Here, the aim is to sell an ideal of personal, individual service that you can buy with ease and enjoy the comforts of the clinical family. Privatization becomes the new national cure. If you want to be an Israeli today you must be able to buy. The state has depended on such a system and even fostered it in order to establish the State of Israel as a major global player, keeping itself financially viable.

> Just arrived back in Israel. Apparently a car exploded last night near my old flat in Jerusalem.
>
> I'm hanging out on the rooftop terrace at my aunt and uncle's house. My cousin and his friend Yaniv are smoking joints and have loud techno/trance music playing. Yaniv fought in Jenin – I think during the recent IDF massacre of the town. I ask him what he saw there. He says, 'there was shooting'. Then he is silent. Later, I ask him again, I'm curious, and not just for the obvious reason of wanting a first-hand account from an Israeli soldier. It turns out he's decided to leave the country for a few years. Go to Los Angeles 'to chill out for a bit'. What did he witness in Jenin? What did he do, I wonder. I ask him if he saw a lot of dead people – he says 'some, not a lot'. I'm not sure I believe him. I'm not sure he believes himself! (Fieldnotes, 30 May 2002)

The idea of the 'embryo method' is a rich resource that demonstrates how embryos get mobilized by privatized medicine for nationalized goals of improvement of the self and the state through the financial success of this private IVF business.

The image of the dead embryo in the clinic, however, complicates the message. As an image of the positive, global capitalism in which the clinic is a player and the patient a beneficiary, the vehicle for transmitting the idea, the dead embryo, brings us once again to the interminable uncanniness of reproduction being about death, as per Foucault and Agamben.[12] In Israel it is about the constant threat of 'the terrorist', dying in bombings or in deployment as soldiers; this combined with the fear of familial experiences of Nazi camps, which are every day being flagged by the state as a constant, global, threat to the Jews.

'Embryos are our baby'

Figure 3.2 is a photograph that I took of one of the many posters and boards on the clinics' walls. The board is lined with photographs of

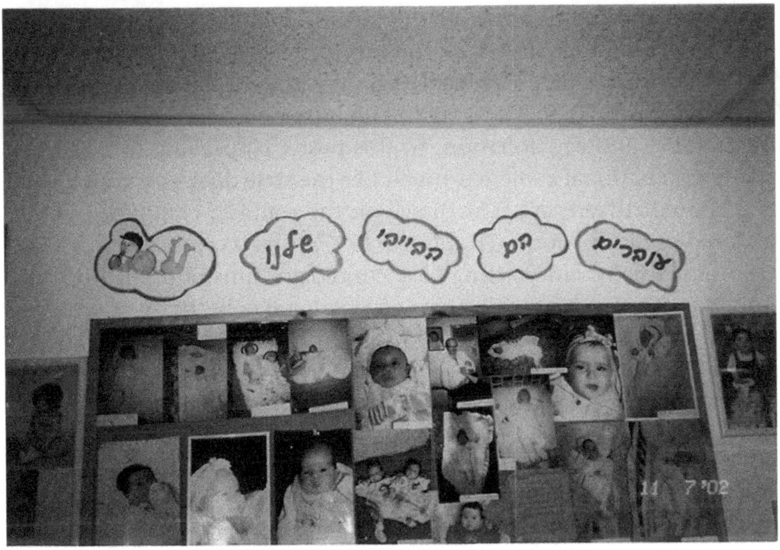

Figure 3.2 'Embryos are our baby' (author's translation from Hebrew)
Source: The author.

babies, including one photograph of a physician holding a baby. One of the items on the board is a letter written to the clinic by a happy couple. The writing above the collage of photos translates from Hebrew as: 'Embryos are our baby'. I wondered, 'how can embryos be a baby?' Who is the 'us' implied in 'our'? Why is the English word 'baby' (in Hebrew script) used in the Hebrew sentence? How does the drawing of a baby next to this writing (at the top left of the image) relate to the statement, 'Embryos are our baby'? What is the relationship between the statement, the drawing of a baby and the photographs of 'actual' babies in Figure 3.2? The viewer is perhaps intended to imagine a direct link between the statement, the drawing of an imaginary baby (wearing a baseball cap, sucking on a 'dummy' and carrying what can either be described as a baseball or a hand grenade), and the photographs of actual children.

Condensed in this poster are fundamental questions of what Haraway (2003) calls, 'kin and kind', relationality and nationalism. The boundaries of the nation are discursively materialized in the poster. The slogan 'Embryos are our baby' seems to be about hoped-for babies, imagined selves and enterprise biomedicine. The clinic builds on clients' desire for a baby through the 'promise' that embryos will result in babies.

Jewish Israelis are not afraid of the idea of embryos or of technology making babies. In fact use of technology coheres with the Jewish ethos of reproduction, where the 'goal sanctifies the means' (Prainsack and Hashiloni-Dolev, 2009; Kahn, 2000). But it is not only Jews participating in Israeli IVF and egg donation. So this poster is speaking to a particularly Jewish national audience, much like the state does – assuming who its preferred citizens are. The image on the wall is a complex, hybrid, 'material-semiotic node' (Haraway, 1997) that connects individual desires for a baby, enterprising biotechnology's aims of profit-through-promise, with the desire to be part of a national collectivity. It provides a sense for the way in which Israeli nationalism merges with global economic concerns, through the way that the image collapses time and how babies become 'figures'. Donors' and recipients' labour are outside the frame.

The image emphasizes the clinic's ability to purvey embryos swiftly to customers, but mobilizes a uniquely Jewish sense of time and hope. The Jewish visualizing of Messianic time and the ideas entailed within it are framed in ethnic/nationalistic tropes. *It is this particular mix of ethnic nationalism and local Israeli neoliberalism that make the image interesting.* Judging by its prominent placing in the centre of the waiting room of the central clinic of this network of ova donation clinics, it can be seen as one way in which the clinic promotes imaginaries of hope for the procurement of *ova* and *embryos* rather than for a *baby*. It presents a narrative of scientific practices of reproduction, faith in technology. At the same time it unabashedly presents itself as being the 'embryo business'. Whilst such a self-representation might seem crass in a British context, in an Israeli one to say that you are a leader in buying and selling ova might not be seen as morally and ethically reprehensible. Although there is an ethical debate in the Israeli government and feminist organizations around payment for eggs, the Egg Donation Law of 2010 permitted this payment. This is not the case in Britain and parts of Western Europe. There is reason to suggest that what is actually being evoked in the image above, as in other practices of this clinic, is a shift towards privatized medical practices in which globalization and Americanization are standing in for deregulation and marketization-type thinking. This company was at the time also engaging in a policy front to legalize payment for eggs. Yet at the same time this is happening through national imaginaries, which are specifically connected to Israeli ideas about what is ethical, what is preferable. Some might suggest that this payment 'proves' that Israel has been Americanized (Avraham and First, 2003). American monetary investment in the Israeli economy is crucial to bear in mind here (Nitzan and Bichler, 2002). Whilst others would claim that this is more a kind of 'globalization' (Liebes, 2003). But it is not possible to separate out the 'local' and 'global' (Lambek, 2011). Israel is positioned within the global economy as an outpost of Western imperialism in the Middle East and has its own national project of survival based on ideologies as well as histories of Jewish victimization. This national-global project is one in which the State of Israel has always been enmeshed.

The clinic's making of embryos and babies into a singular thing may be about a national hope that the daily actions of people will result in a desirable outcome (see also Morgan, 2009). The women from whom the eggs are derived are profoundly absent here. Eggs without a mother for mothers without eggs, this is at once a tragic and complex relationship among women, a convenient transnational business relationship, and it says something specific about the way in which Israeli nationalist thinking about land and occupation translate into the realms of reproduction. The embodied subjects that we saw in Chapter 2 are absent from this narrative other than as viable sources of an important resource. Also the women who carried the fertilized ova to term and made babies with their bodies are not present (most likely because they wanted to conceal their identities). The photos are of babies, or figures that stand for something else: triumph against adversity, reproduction through highly technical means, transnational eggs of the right kind.

Embryos are 'promissory' (Thompson, 2005), and involved in an economy of reproduction rather than production (or capital accumulation).[13] Thompson's work is based on research that she conducted in IVF clinics in the USA. As part of her study she examines the kind of capital that gametes embody. She states that an earlier version of capitalism involved measurable output and productivity, yet in relation to biotechnology and reproduction, new forms are emerging. According to Thompson, in the world of biotechnology, capital is not 'accumulated'. Yet in the Israeli clinic that is under examination here, there is accumulation of capital resulting from the selling of embryos. In contrast to Thompson's notion of the promissory then, in privatized Israeli medicine the capital of eggs is both promissory (there may or may not be eggs, embryos and babies) and accumulative (the clinic will profit). Biotechnology and reproduction are made into capital through the 'promise' that they offer about their future success. After all, Israelis can go to public clinics and have their IVF treatments fully covered by the state (but recall from previous chapters that egg donation was at a standstill in the country so that they felt that they had no other choice but to go private). Thus, the existence and continued profitability of biotechnology depends on its ability to be taken seriously as a 'stock' or 'future'.

Similarly, the statement 'Embryos are our baby' depends on the person viewing it believing in the idea of an embryo as a 'future' baby. What constitutes the 'promise' of embryos is their successful transformation into a signifying resource for the clinic, which results in the clinic's accumulation of capital. Whilst they are promissory for recipient couples, embryos are accumulative for the clinic. What is crucial here is how all of this furthers a national agenda of hope for survival. And now that the welfare state is not functioning in its idealized way, the hope and capital are invested in private companies. Yet despite this taking reproduction away from national interests through privatization, it is still profoundly nationalist.

Abridgement, Messianic time

Taking this argument further, I want to suggest that in this image's slippage between embryos, babies and hand-drawn images of babies, there is a skipping of *time*. Taken as a whole, this image – the statement 'Embryos are our baby', the juxtaposed photographs of babies and the drawing of a baby – does not differentiate between embryos and babies, implying that egg donation will happen quickly at this clinic, whereas at other clinics people often wait for several years. The urgency of the company's brochure shows that this clinic is delivering a fast service to their customers, but this has a specificity in Israel in relation to the way in which racism works there (Chapter 4) and its pronatalism – and how all of this ties in with violence in the present and the past, a particular nexus of the urgency to populate. Embryos are Our Babies and the Embryo Method, as specified in the brochure, abridge time.

According to Benjamin, for Jews, every second of time – past, present and future – was a possible moment when the Messiah might arrive (1999 [1947]: 255). For him, every second of time was filled with hope and anticipation of salvation. This idea of hope is built into Benjamin's notion of abridgement. This religious and mystical notion of time is in-built into the Israeli state's ideal of hope, that rather than being about the coming of a messiah it is about the survival of the state. This abridgement is perhaps one way in which the clinic naturalizes the relationship between embryos and babies. It suggests that embryos always result in babies, as opposed to the many cases in which transferred embryos do not grow into babies.

It defies the narratives of the women that I interviewed in Israel about ova donation, where overwhelmingly embryos resolutely did not become babies. From my observations, it seems evident that the clinic puts much effort into managing the expectations and emotions of their patients. It does so in many ways, one of which is by encouraging attention to embryos.

The clinic may also be reproducing Jewishness through the mobilization of this particular kind of messianic (time-abridging) hope. There are echoes of this reproduction of Jewishness in the clinic in what Dr Shmuel himself had to say. On one occasion he stated that:

> Three years ago, well, more than three years ago, but three years ago it started to work 'full volume'. And *we saw it was good*. There were many legal *hurdles*, but we received authorization from the Ministry of Health [in February 2002]. [...] So we said, we'll take the sperm there and bring embryos *la-Aretz*. We checked [the legality of it] and it turns out that at the moment that embryos are formed they belong to the couple in treatment. The Ministry of Health told us there was no legal obstruction. Because, you know, there's no such thing as *'ubbarim mesoravei aliyah* [embryos who've had their 'right of ascent' refused]. (Dr Shmuel, interview, Israel, 11 July 2002)

Dr Shmuel states that when he first opened his clinic, he 'saw that it was good'. This is a direct quotation from a biblical passage (from the first chapter in Genesis in which God is represented as creating the day and the night, animals and so on), which injects a religious element into the privatized medical practice of IVF at this clinic. It is common in Israeli Hebrew to insert biblical phrases into everyday speech (see Dominguez, 1989; Boyarin, 1996). However, the quotation is interestingly placed here. Judaism is a 'civil religion' (Liebman and Don-Yehiya, 1983), its presence is felt throughout public events and the state uses Jewish religious symbols to strengthen non-religious policies. In contrast, in the USA religion is generally kept to the private sphere. It might be imagined then that this private space of the clinic with its connections to the USA and Europe was perhaps less likely to be imbued with religiosity. However, the presence of biblical quotation renders it possible to see the clinic as something other than 'merely' a secular, private medical space where ova donation occurs, but also as a Jewish space, despite the non-Jews who access their services.

Furthermore, the context of this particular quotation, coming at the end of seven days of God's creation indicates that the clinician may be playing God himself (see also Roberts, 2012). The assertion that this was done despite legal hurdles places his endeavour within a narrative of 'scientific progress' *through* struggle (Franklin, 1997). Its juxtaposition with biblical quotation places it in an ethnic-national narrative of struggle. It is through *struggle* that Israel is narrated as having achieved its status as a nation-state. This intersection of the national with the scientific continues most pointedly when he argues that there is no such thing as an embryo that has lost its right of 'ascent to Israel' (*aliyah*). This is one moment where the religious, the national and the scientific collide. This notion of *aliyah* is juridically associated with a notion of 'return'. The 'Law of Return' guarantees the legal right of Jews to 'ascend' to Israel. It is encoded into the legal framework of the state; in this sense, it indicates the Israeli narrative of a 'return', after a 2000-year absence. Dr Shmuel is thus directly speaking about citizenship. In the Law of Return, an historical link is made between Jews living in Israel today and Jews who lived in the Land of Israel in biblical times.

This is the kind of 'simultaneity' that Benedict Anderson discusses when he describes how nations are imagined (Anderson, 1991: 24). The imagined past is made concrete by being inscribed into the legal framework of the state in the present. And the specificity to Jewish ideas of simultaneity must not be missed here. Annually on the holiday of Passover it is customary to read the Hagaddah, even among non-religiously observant Jews, where we are all called (interpellated) into the collective by having to imagine yourself as if you received the Ten Commandments from God on Mount Sinai. Synecdoche is central to Jewishness – each Jew, in every period of time is supposed to have been present at Mount Sinai receiving the word of God.

It is indeed true that embryos cannot lose their right of ascent, as Shmuel indicated in the passage quoted above, due to the fact that they are not 'persons' under Israeli law. The fact that it was 'Jewish' or Israeli sperm that left is what allowed the fertilized ova to be constituted as 'returned'. Lawyers and representatives of the Israeli Ministry of Health all referred to this notion of the 'Right of Return' of embryos. They spoke in this fashion because of a court injunction preventing the import of ova from Romania where lawyers took the approach of the right to 'return' embryos to their commissioning couples (this case took place in the months just prior to my research at Shmuel's clinic and was already resolved at the time of my speaking with Dr Shmuel). It was legally decided in this case that embryos did not indeed constitute persons. Yet the language of the case against him 'stuck'. The embryos are, on the one hand, labelled by Shmuel as belonging to the commissioning ova recipients and, on the other hand, to the nation. In part, this facility to speak about embryos as having or lacking a 'right of return' also makes it possible to speak of them as being, at least symbolically, a 'baby'. This seems a unique instance in the global scene of IVF in which the legal system and doctors discuss and adjudicate citizenship based on rights of return to biogenetic substance. Whilst we have seen in other contexts such as Egypt the negotiation of religious acceptability of such substance, the Israeli state's unique citizenship laws make it stand out. Here it is also about acceptable future citizens.[14]

The local neoliberal: choice, variety and the hope for donor/embryo/child traits

One particularly noteworthy aspect of the 'embryo method' is the way in which it promotes an Israeli neoliberal sense of 'choice'. It has been argued that the notion of 'choice' operates in a problematic way in IVF, emerging as it does from a market-oriented notion of freedom (McNeil et al., 1990; Strathern, 1992b). Indeed, in some cases women can be almost coerced into going for IVF because they feel that they have 'no choice' (Franklin, 1997). In Israel the very proliferation of choice indicates the rapidly increasing consumer culture – sometimes enabled by the way in which the Israeli market has shifted towards and sometimes been constrained by Jewish religious or cultural factors (Carmeli and Applebaum, 2004).[15] So, for example, many couples are not able to choose to marry outside the religious system but instead choose to fly to places such as Cyprus to have civil ceremonies. As well, abortions are not encouraged for Jewish women especially, except in cases where foetal anomaly is found in screening (Weiss, 1994).[16]

Whilst it is tempting to view the flows of people and bodies across borders as free and unfettered, it is my contention here that in imaginaries of Israeliness in egg donation, the borders of national bodies are only semi-porous. Whilst a sense of cosmopolitanism is espoused as official discourse in Israel, this is often limited to consumption of food, art, travel and popular culture. To become actually a '*frenkit*', '*chach-chach*', '*'arss*' (common derogatory terms used by Ashkenazim about Mizrahim) or Arab (i.e. the Other) is generally reviled within this culture despite some examples of European Jews who have 'gone native'

(Shohat, 1989; Motzafi-Haller, 2004). What can 'safely' be incorporated into the Jewish Israeli body as acceptable is very particular. It is important to see that the desire for a child has a racialized geography, in the sense that desired ova travel to Israel from Europe. Romania, on the one hand, is imagined as a source of possible pollution to the national body and, on the other hand, as a source of desirable traits.

Soon after Dr Shmuel's transnational egg donation programme was up and running, news stories regarding the imagined threat of HIV infection of these 'foreign' eggs began to spread in newspaper reports about ova donation (Freilin, 2000). I learned from interviews with members of the Israeli Ministry of Health, the committee that legislated egg donation and Dr Shmuel's own company lawyer that this media scare led to a temporary ban on the importation of ova from an Israeli clinic in Romania, where I conducted part of my study. The clinics that were running this transnational egg-donation programme and the women accessing care at these clinics soon mounted a case to the Israeli High Court of Justice to eliminate the ban on importing the ova. The lawyer representing the case referred to the situation as a 'state of emergency' for egg recipients waiting for eggs from Romania. The prospective ova recipients whom I interviewed were portrayed by the media, themselves and healthcare workers as being in 'need'. Indeed, one ova recipient whom I interviewed at a private clinic said that she was waging a 'war to make a baby' (Chapter 4). The strategy taken by the clinic, of petitioning women needing ova and of framing them as victims of infertility needing ova was successful. The case at the High Court of Justice was won, and women in Israel continued once more to receive ova from Romania.[17] These are moments in the policy struggle that led to the present-day Egg Donation Law, 'liberalizing' the practice to attend to the fertility needs of Israeli women. There were as well many hurdles in getting this legislated (see Chapter 6).

The flip side of this sense of Romanian women as potentially 'polluting' to Israeli women and to the nation is that for most Israelis whom I interviewed, Romanian donors were seen as perfectly acceptable.[18] When I spoke to Dr Shmuel about this acceptability, he said:

> Look, in Romania the population is very similar to the Israeli one. There are dark ones, there are blonde ones there are black ones. There's everything you want. Don't forget that there were two large waves of immigration (to Israel) from there, one of which was from Romania [...] so the population of Israel, not the ones who have recently arrived, and the population there are very similar. However

> the main reason for going to Romania was the availability. To get eggs and not keep a 49-year-old woman waiting for five years. Three months' wait and she has good quality eggs from which she can become pregnant. (Dr Shmuel, interview, Israel, 11 July 2002)

Ethnicity, nationalism, immigration and availability are brought together here. Genetics and disease testing are mobilized as narratives for avoiding 'pollution'. The availability of Romanian women as an inexpensive, young and racially desirable resource is the promise offered by the 'embryo method'.

Despite the model of privatized medicine as global at this clinic, the reproduction of Israelis and consequently of the *Jewish* nation is still what is implied here. A Europeanized Jewish body is envisioned as the ideal. This is best exemplified in the 'Preferred External Features' form. After their initial 'intake' meeting, ova recipients are asked to fill out a form labelled, 'Preferred External Features' (see Figure 3.3). I analyzed 32 of these forms. This form enlists patients as 'choosing subjects'. Through choosing traits of donors, ova recipients imagine what their hoped-for donor might be like and what kind of baby they would like to have.

I found that 26 out of 32 (81 per cent) women preferred light-skinned donors; 11 out of 32 (34 per cent) preferred brown hair; 4 out of 32 (12.5 per cent) preferred blonde hair; 10 out of 32 (31 per cent) preferred blue/green eyes; and 6 out of 32 (19 per cent) preferred brown eyes. These numbers and percentages would be meaningless if intended to be representative. However they are indicative of how people imagine their egg donor, their intended baby and their commissioned embryos. (Chapter 4 analyzes some of the comments that patients wrote at the bottom of these forms). The selections indicate the desire for light skin; this is the main distinguishing aspect of people's selections. This can be seen as a desire for 'whiteness', the idealized appearance in the West (Dyer, 2000; Frankenburg, 2000). In the Israeli case, such a desire links up very strongly with the ways in which Ashkenaziness (or European Jewishness) still dominates as a site for cultural emulation. A long-documented fact in Israeli society is the way in which non-Ashkenazim were treated as more primitive, requiring civilizing, and are discriminated against in Israeli society. The cultural roots of contemporary Zionism as a political movement emerging from Europe and driven by Europeans must not be forgotten here. These facts clearly influence the choices of individuals on the 'Preferred External Features' form.

Company logo

Photo of recipient

Preferred External Features

Mark an X next to the donor features you prefer

Date: ______________________________

Name of Patient: ______________________

Skin Colour:

Light	Brownish	Dark	Very dark

Height:

1.6	1.7	1.8

Hair Colour:

Blonde	Brown	Black	Ginger

Eyes:

Blue/Green	Brown	Dark

Comments: ________________________________

____

____

Company Name and Contact Details

Figure 3.3 'Preferred External Features' form (author's translation from Hebrew)
Source: UFC Clinic, Israel.

As well as showing a preponderance of desire for whiteness (and this is a much more lengthy discussion that I enter into in Chapter 4), the documents that I analyzed are a technology of the clinic (Howell, 1996). The 'Preferred External Features' form seems to elicit 'hope' from the recipients. It asks them to project their expectation for a baby into the delineation of physical types for would-be donors, linked to particular phenotypic traits such as hair colour and height. The form, in effect, asks recipients to imagine that traits can be *selected*, and requires that they *believe* that such traits can actually be chosen. The percentages cited above are only indications that the clients of this clinic believed in the technology of the form; that it would get them what they desired.

The head of the clinic, Orit, conveyed to me the idea that there is some sort of systematic pairing of recipient choices with donor traits. At our very first meeting, Orit told me: 'The eggs at this clinic are brought in from a clinic in Romania [where] they are matched according to eye colour, hair, body structure, height, skin colour [...] we *meshadkhim* [broker a marriage or negotiate an agreement] between them and the recipients.' It seems that the clinic's staff members collect the forms and enter the details therein into a computer spreadsheet program, and then the forms are put away in a filing cabinet. In practice, however, the clinic does not seem to follow up on the 'choices' made by recipients. There is no systematic matching of recipients' choices with donor traits either in Israel or at the clinic in Romania where the ova are fertilized, or at least I did not observe any kind of 'matching'. An embryologist at the clinic told me that Sasha, the chief embryologist, matches the donor traits with the preferences expressed by couples. A few weeks later I spoke with Sasha and asked him how this 'matching' is done. He replied, 'What matching? There's nothing like that!' A few months later when I went to the clinic in Romania where ova are donated, I noticed that Sasha would list donor eye and hair colour in a spreadsheet program on a computer in the laboratory. I asked him again about matching and he did not change his answer. On one occasion as I watched data being entered, after having seen the donor myself, I noticed incorrect eye and/or hair colour being entered into the database.

The fact that no one actually matches recipients' requests with donor traits makes it even more apparent that the 'Preferred External Features' form is a technology of engendering hope. It telescopes donors into 'donor traits' on the 'Preferred External Features' form, and again condenses embryos, traits and donors with imagined babies. This seems to be a common practice of both ova donation and sperm donation in other systems of privatized medicine (Birenbaum-Carmeli and Carmeli,

2002). The process of 'selection' is interesting as it provides a different sort of narrative about biological reproduction of traits to the typical scientific one, which is about the complex biogenetic processes that determine physical features in people. The discourse here gives a sense that it is indeed possible to 'choose' a blue-eyed baby, with an expected height of 1.7 metres. It allows people to invest their hopes in science's 'ability' to 'know' what will happen.

These forms enact 'new assemblages' (Rabinow, 1999) of Israeliness, by bringing together consumer choice, national bodies with transnational reproduction. By reinscribing desirability for particular phenotypic features they also perform 'old assemblages' of Israeliness (by this I am referring to the State of Israel's history of exclusion, racism and the reproduction of whiteness; see Nahman, 2006). While the 'national' is literally reproduced through 'transnational' exchanges of ova, the 'national' is also reproduced as Ashkenazi.

Dr Shmuel was interested in making his business the most successful egg providing clinic in the country. Indeed everyone that I interviewed in Israel knew about it, had been to it or wanted to go to it. His clinic was vetted by the Ministry of Health, so it had connections with the state despite being a private clinic; while representatives of the ministry told me of the immense problem of the egg shortage in Israel, and at least some of the National Health Funds paid in part for treatments leading up to egg donation. But questions remained on the diminishing relationship between the welfare state and health, which had traditionally been available to all (Filc, 2009). At the time of the research, the *kupot* paid for some of the hormones.

I am having a chat with Orit, the head manager of the private egg donation clinic I've been hanging out in for the past few months. She says that *Kupat Holim Maccabee* [Maccabee Sick Fund, one of five national healthcare providers in Israel] will now start reimbursing 5000NIS [New Israeli Shekels, about US$1063] out of the total cost of 17,000NIS [about US$3617] for ova donation. The rest of the *kupot* have higher subsidies but still not that much. But, Orit adds, she disagrees with the fact that the government pays for IVF. She doesn't believe that the government should cover it, 'even if the government wants more people, because afterwards many go off to be soldiers so we need some sort of (demographic) support'. (Fieldnotes, 2 July 2002)

Unsurprisingly oocyte recipients and prospective recipients tended to be much more in favour of the government-funded healthcare providers financially reimbursing egg donation, even when it was done at a private clinic. In having a baby and going to a private clinic for the best (quickest) treatment possible, they were doing something for the national good (see also Nahman, 2011). Shakhar, the male partner of an egg recipient interviewed at Shmuel's clinic said:

> So, can't the state make a law that helps or that pays [...] why not take tax-payers' money to help families who fall into this situation of having difficulty bringing children into the world. Is this trivial? Let's say Hanna [his wife] and I are in this situation today where we have to pay for the treatment out of our own pockets. A day will come when my son asks me, 'Dad, should I go to the army?' I'll say to him, 'If you can help it, don't go! *Do for the state as it did for you*!' (Shakhar, interview, Israel, 25 July 2002)

A heated discussion ensues between him and his partner Hanna:

Shakhar: What's the next step? That sick people in need of a blood transfusion will have to pay for it? For how long can the *kupot holim* play with this issue? If it's only eggs, then they'll get away with it. Or cancer, if they tell a man who is dying with cancer that they will not pay for his treatment, he knows that either he pays or he dies.

Hanna: My feeling is that a couple who has money and is in this situation will most likely have children. A couple who hasn't got money, then the apathy and lack of empathy in the *kupat holim* states that: 'Okay, you don't have money? Then you won't have kids.' (Shakhar and Hanna, interview, Israel, 25 July 2002)

The demand that the *kupot* care for and empathize with people in their situation actively knits the state to its citizens. Conversely, should the state not take on the role of parental caregiver, the citizen should divest her or himself of responsibility, according to Hanna. They are negotiating a contract of mutual care here.

Hamid, a man who self-identifies as Palestinian, states that:

> Two years ago we started treatments again and have undergone a series of abuses with the *kupat holim* because they said that people with adopted children do not have the right to receive treatment. [...]

> They say the medicine costs 160 shekels and you pay only 20 shekels. The injection ampules alone cost 2500 shekels, add to that the pills and the Decapeptil, which is another 330 and the Chorigon and you get to 3000 shekels. The *kupa* said I should pay the 13,000 up front to cover the whole thing and I said I can't give this money all in one shot, I'm not Strauss and not Elite.[19] So the nurse made some phone calls in order for us not to have to pay this upfront fee. (Hamid, interview, Israel, 3 July 2002)

The healthcare providers are often called upon to be caring intermediaries between the patient and the supposedly benevolent state.

This set of remarks comes out of a wider public debate and anger at the loss of the welfare state in Israel. There seems to be a hope (*tikva*, echoing the name of the Israeli national anthem *Ha-Tikva*) that some kind of benevolent state can be recuperated, with enough dissent and enough protest. The recent J14 movement in Israel involving 'tent cities' that sprouted up across Israel in the summer of 2011, sparking some of the largest demonstrations in Israeli history, bear testimony to the widespread anger at the government for backing down from the paternalistic claims that it has been making since its inception. If the Israeli state is a family, the claim of the citizens goes, then the government should support its people. One aspect of these demonstrations and occupations is the way in which people who tried to bring in to the discussion the question of the occupation of Palestine and the racism in Israel were virtually silenced. The organizers of the tent cities were adamant that the focus remain on economic issues and not on border, war, occupation and race issues. Mizrahim and Palestinian citizens of Israel managed to make themselves heard within these arenas, the effect seemed to be some level of education, but little changed in the J14 ethos. And the tent cities dissolved, although the anger did not dissipate. Becoming an egg recipient also involves becoming acutely aware of where you stand in the economic order and in the government's list of priorities, according to my informants.

Conclusion

This chapter traces some of the discursive practices of a private egg donation clinic in Israel, seeing how private medicine fosters national concerns, materializing the state, and how this is done transnationally. The elaborate and sometimes illicit procurement of body parts has received significant attention from anthropologists looking at organ trafficking. Such accounts are compelling insights into how globalization, medicine, free markets and individual illness mix. In Israeli ova donation there was a 'reverse traffic' at play: it was not trafficking of eggs in the typical sense of illegal transport. Indeed the doctor and other clinicians were extremely candid about how and where they receive their eggs. The Israeli Ministry of Health also approved the process. Yet in its descriptions of the method of acquiring eggs and making 'embryos', this clinic focuses the attention of their customers on the beauty, youth, willingness and availability of their donors, and on quick service. There is little appreciation for the exploitation of egg sellers that is discussed in Chapter 2. By reversing the process of travel, doctors going to Romania with sperm and returning with fertilized ova, the conventional definition of trafficking is circumvented. The focus can be much more on the ease of this service, helping Israelis to have babies.

The chapter demonstrates how the private sphere in Israel serves the larger nationalist agenda. The clinic tries to funnel peoples' hopes for a baby into hopes for embryos, which is what they specialize in and sell. Through imagery, branding, logos and wall posters, the clinic subtly and unsubtly reinforces this process. Through an analysis of a set of images, discourses and documents employed by this private clinic, this chapter shows how rationalized notions of speed and efficiency combine with deregulation and market 'liberalization' and are evoked through racializing tropes of availability and selection, and are enmeshed with Jewish and Zionist messianic hopes for self-improvement. That the terrorist and violence of extraction (of the desire for the removal of Arabs from Israel) lies in the midst of this makes it increasingly difficult to provide a cleansed, distanced narrative of Israeli reproduction.

Ironically, the reverse traffic sometimes happens through actual description of the 'process' by the clinicians through images, brands and narratives. These imaginaries of the body are imaginaries of the nation, offering 'infertile' Israelis the hope for self (national)-improvement; what Lawrence Cohen (2011) has called 'supplementarity' in relation to organ donations. They are messianic 'technologies of hope' for 'choosing' ideal nationhood through selecting donor/embryo/baby traits. At stake here are the *kinds* of bodily imaginaries that are reproduced through the seemingly contradictory prosthetic possibilities offered by reproductive technologies at this clinic: *transnational, privatized and nationalist*. The imaginaries materialize the individual and the state (Nahman, 2006), both are made concrete. By examining the hope invested in ova and embryos in the discursive practices and images in one private IVF clinic in Israel, I argue that 'the nation' is made through transnational privatized neoliberal practices and discourses. In this account, a hybridized version of Israeliness emerges, one that mixes Zionism with transnationalism and privatized medicine, within the logic of accumulative market ideals.

The Israeli national anthem *Ha-Tikva* as previously mentioned means 'The Hope'. It has a deep, mournful melancholy borrowed, derived, extracted, from earlier European tunes and songs. Its words evoke that 'we have not lost our hope of 2000 years to be a free nation in our land'. That such a hope is promulgated by governments for the building of nationalism is perhaps common to many states. That we find this religious nationalist hope in a private, global space, tied intimately to national violence of extraction is perhaps more surprising. While ova donation at this clinic is practised through a global market of egg donation, some aspects of the material-semiotic nation are entrenched through this very transnational trade. The practices and self-representations of the clinic materialize aspects of 'the national' that are both 'old', in the sense that they resonate with Zionism's ideals of the 'ingathering of exiles', multiculturalism, racism, relatedness and religion; and 'new' in the sense that they involve privatization, neoliberalism and hyper-consumerism. The possibility of selecting desired traits also turns individuals who are considered 'illegitimate' subjects, by virtue of their inability to have children, into legitimate subjects. The abridgement of embryo and baby in the phrase 'Embryos are our baby' may also be one discursive node through which a sense of 'chosenness' is symbolically imparted to ova recipients in Israel. Since embryos may literally be the only offspring that recipient couples will have, the abridgement of embryos and babies may enable the recipients to become legitimate reproductive and therefore national subjects. The clinic's wall poster 'Embryos are our baby' presents the promise of redemption of the 'unchosen' body *through* a project of transnational trade and consumption. Recipients are encouraged to regain what they apparently lack in terms of the national idealized self and body (the ability to be parents and therefore proper 'chosen' citizens), through a consumer imaginary, whereby appropriate citizenship can be bought in the form of embryos. This is the ultimate 'retail therapy'.

This condensing, in imagery and practice, shifts attention away from the effectiveness of scientific technique toward *intended* effectiveness. In terms of its particular significance, this example indicates how the privatization of Israeli medicine facilitates a global assemblage of nations, bodies and science (Ong and Collier, 2005). *This* assemblage, of 'Embryos are our baby', 'the embryo method' and the 'Preferred External Features' form, mixes Zionism with hope and consumerism. It is a kind of messianic consumer economy that offers the 'promise' of a child, and of racial improvement as well as that of becoming a proper subject through believing in the redemptive possibilities of a privatized, specialized 'method'.

4
Repro-Migrants

> I've been travelling almost all over the world, really. And I am a very open person [...] I am a very friendly person, I am not afraid to speak any language that I don't know. I find my way. I did not want to go to Bucharest; I just did not want to go. I had to go. [...] I wanted to be [...] non-seen [...] invisible [...] like I've never been there.
>
> Dorit, Ashkenazi Jewish Israeli, interview, Israel, 6 March 2002

Transnational trade in human eggs has led to many moral and ethical debates around the mode of these exchanges and their definitions. The experiences and process of becoming an egg recipient, personal histories of migration, race, borders and 'repro-migrations' are the focus of this chapter. The desire for discretion in the pursuit of egg donation is well documented.[1] But the plea of the woman quoted above, to be invisible, was also about not having to go through the egg donation at all. The Israeli women whom I interviewed did not want to have to go through what they often referred to euphemistically as 'this thing'. The technological availability meant that, *eyn berira* ('there's no choice', which in Israel is often used with reference to military and security measures; Handelman, 2004). In Israel I got the sense that one is impelled to use technology if it is available. And because of a lack of available ova 'back home', women felt compelled to travel or buy eggs imported transnationally. Their own personal histories of migration to Israel and the racial politics of their settlement and 'inclusion' into the collectivity are enmeshed with the desire for a child and the contemporary narratives of border defence. Border defence and egg recipient narratives are not always or necessarily linked, this connection is an ethnographic interference. These 'synecdochical connections' enable a more critical account of the wider social context in which all of this was taking place.

Part I: Becoming repro-migrants – pain, travel and motherhood

> My name is Sharona, I'm 46 years old, single, I want to be a mother. It's very important to me to raise a family, a home. Very simply I feel emptiness from not having a child. I feel very bad. Simply I am afraid to be left alone, this is very emotional. So I decided to bring a child to this world. [...] So that I will feel that I am a mother. And I have a right. I want to fulfil that right. And [crying] I want to have a child, of my own, a house of my own. I was never married, I thought to have a child, to meet a man, so that there will be a father for my child but that didn't work out. So very simply the time is running out. I don't want to miss [...]. I've about two or three years left. I'm scared.
>
> I've been having treatments for a year. The *kupa* stopped paying for my treatments when I reached 45 years old. (Sharona, Kurdish-Jewish Israeli, interview, Israel, 29 August 2002)

Sarit: When we got married there was this problem *ba-Aretz* ['in the Land', shorthand for the State of Israel], that you couldn't get an egg donation here. So we did our first treatment in Prague. We couldn't wait for the eggs here, there just weren't any, and time is of such importance to us, we are not so young.

Yokhanan: As soon as we were married we consulted a professor at the [private hospital]. And there we went through a process of about 1.5 years, back and forth [...] a series of tests from here to eternity, always with the feeling that something would happen imminently, and nothing happened. With all due respect to the professor, he was a very big disappointment. He turned us into the mill for his personal savings programme. Each month we paid him 600 shekels for treatments we could have received at the *kupat holim* in an orderly manner. He always told us to come back for another test, and we didn't know, we didn't know, we were complete ignoramuses. And

> we were sent by our GP to this professor, we thought we had reached the top doctor in the country for egg donation. [...] And I finally understood that the professor was simply dragging us along and there was no outcome in sight, because after a year-and-a-half of tests we still hadn't done one full cycle of IVF. And he kept telling us we are waiting, waiting for eggs, and they will come soon. But the fact is the eggs didn't come. And this really hurt us. So we went to another professor, this time at a state hospital in Jerusalem and he sent us to the clinic in Prague, where the head of the clinic is Israeli. I don't remember his name, but we could speak Hebrew with him, and we easily made the connection with him. We went there, to Prague, and left the sperm there and he told us that within half a year we would get a donation. And within one year exactly we had the first fertilization. Exactly one year. And this whole thing cost us US$3500, not including flights and micromanipulation [techniques involved in creating an embryo under the microscope; see Chapter 5], in short, a respectable amount of money that for which the *kupa* does not give any refund. This is a disgrace to the state that they treat women who need egg donation as *sug bet* [type B or second class] persons! And what do people who can't afford this do? We are cutting down on our expenses. Everything I own is now part of the 'fund to bring a child to the world', that's how I call it [...]. And my heart goes out to those couples who don't have any money. The treatments, the internal tension is terrible, and the lack of calm [...] and on top of this you add the cost. (Sarit and Yokhanan, Jewish Israelis, interview, Israel, 25 June 2002)

The process of becoming an ova recipient involves decisions about whether one would accept ova from another woman, and strategies of negotiating the importance of genetics, the relevance of gestation of the ova and how aspects of a kind of biological thinking come in and out of recipients' ideas about having a child through ova donation. I have chosen to begin by considering the very complex and personal processes of becoming an ova recipient in contemporary Israel. This section is intended to indicate the ambivalence of ova recipients with regard to their use of ova donation as a way of becoming pregnant.[2] As this chapter demonstrates, the 'local biological' (Franklin, 2005) in Israeli egg donation is intimately tied up to the history of Jewish migration to Israel from various parts of the world, including Arab states, Europe, Asia and Africa. Here the 'local biologies' (Lock, 1995) involve histories of travel, reticence towards reproductive travel (what I call 'repro-migrations'), making the question of 'local' versus 'global' a moot one (Lambek, 2011).

An ova recipient is usually a woman who has undergone many IVF cycles that have 'failed'. If these unsuccessful attempts at conceiving a child are deemed to be due to the 'quality' of her own oocytes, a physician might suggest that the woman try ova donation. Sitting with me in her kitchen in a town south of Tel Aviv, the head of the Israeli patient advocacy group CHEN ('Charm', named after the founder's father; although an Israeli anthropologist pointed out to me that as an acronym, CHEN also stands for 'Cheil Nashim', the Women's Corps in the IDF), an ova recipient herself, characterized three 'groups' of ova recipients. The first includes women who have reached an age at which their ova are considered to be 'unviable'. These women are generally over 40 years old. The second category is women who, for reasons that are unexplainable (since they are young and considered otherwise to be 'healthy') have not been able to conceive with their own ova. The third category is women who were born without ova, or who had previously suffered from cancer.[3] These are the main groupings of women in Israeli ova donation programmes (Ofra Balaban, co-founder and head of CHEN, personal communication, February 2002).

Balaban 'mapped' or 'grouped' ova recipients for me, a sociological mapping that was intended to help me 'make sense' of the population. This kind of categorizing thinking pervades Israeli society and the military (Handelman, 2004) and has been a biopolitical tool for managing populations. I perform it here with tongue firmly in cheek in order to enter into a discussion of the politics of race and racism in Israel.

The 25 women whom I interviewed who painstakingly decided to become ova recipients have varied ethnic, religious, class and personal backgrounds. Most of them were born in Israel. Most of them are Jewish. Of those interviewed, 20 per cent (4 interviewees) were with recipients who self-identified as Palestinian, Arab or Druze. It was more difficult to distinguish among ethnic 'groups' in Israel as couples are often 'mixed' and many people consider themselves to be both Mizrahi[4] and Ashkenazi.[5] However, 24 per cent (6 interviewees) overtly self-identified as Mizrahi. The remaining 15 interviews were conducted with people who were either of 'mixed' ethnicity, Ashkenazi or couples in which partners self-identified as coming from different ethnic backgrounds. A large proportion of people called themselves 'mixed', by which they meant that they had parents who are Ashkenazi and Mizrahi. The majority of my interviews with egg recipients, then, come from non-hegemonic subjects, people who have historically been on the margins of Israeli society. Interviewees came from different socioeconomic backgrounds as well. Most of the women lived in central Israel, cities such as Tel Aviv, Herzliya, Rishon Le-Tsiyon, Holon, Jerusalem, Netanya and Petah Tikvah. Most of the 'ethnically minoritized' people, Palestinians, Druze and those who self-identified as Mizrahi came from more northern and southern cities, Palestinian villages in Israel, Jewish settlements in the West Bank and *moshavim* (Israeli cooperative village or settlements; s. *moshav*).[6] These demographics will be important to bear in mind later on in this chapter, they hint towards the questions of mixture and difference that materialize Israeliness (Nahman, 2006).

Attention to the details of ethnicity means undermining the social marginalization and categorization of people whist at the same time attending to and enabling a critique of how gender, race, class and borders are made in Israel (Lavie, 1996, 2011a, 2011b; Motzafi-Haller, 2004; Shohat, 1989; Alcalay, 1993). By putting Mizrahi women at the heart of the interpretive description here, I am attempting to combat the tendency within Israeli society and political sciences to have them/us as a category of analysis rather than (as I intend Mizrahi women to be here) as 'speaking subjects' (and see also Motzafi-Haller, 2004: 178).

Most of my interviews were conducted at the Tel Aviv clinic, either in the patients' recovery ward (after or before 'embryo transfer') or in the clinic's conference room (the small boardroom). This location was slightly problematic as it did not afford total privacy. If I interviewed women by their bedside, I was conscious of the fact that, dressed in hospital 'greens' and sitting by their bed, I resembled a healthcare provider. It was important for them to realize that I was not a member of the staff and that they could talk openly with me about the treatment that they received at the clinic and about their opinions of the clinic. I attempted to compensate for the potential confusion that my dress might have evoked by telling patients a little about myself and my research (including where I came from, what sort of degree I was doing and so on). I made repeated efforts to distinguish myself from any kind of healthcare professional, especially social workers, for whom I was often mistaken.

Interviewees were often recruited by a member of staff who would approach the woman or couple, asking if they were willing to talk with me and introducing me to the interviewee. I would enlist this assistance because I felt that it would be intrusive to impose myself on people during this very sensitive time. I also handed out interviewee recruitment sheets in the clinic, and made them available at the reception desk. Interviews were recorded on a mini disk and cassette recorder. I also kept notes in a notebook. While I was interested in learning about the experiences of women and men who were having IVF with egg donation, women often told me that they wanted to speak with me in order to help other women. I explained to them that, although this research is largely focused on women's experiences, I was not sure how or whether it would help women. Their response to this was to indicate their hope that, nonetheless, others would benefit. I suspected that their concern also facilitated an opportunity to talk about how they felt about their treatment, given that they were rarely provided with such an outlet by the clinics.

In similar ways to egg donation in other social contexts, becoming an ova recipient in Israel involves usually painful decisions about whether to stop trying to have a baby altogether, whether it is important to gestate the pregnancy oneself and how to tell others about this decision. The replies often suggested that age was related to failure to conceive and that even knowing about infertility was destructive. They resonate in some ways with the 'I can't have a baby stories' that are part of a Western tradition of narrativizing motherhood or the difficulties encountered in trying to become a mother (McNeil, 1993a, 1993b). Sometimes the act of telling such stories was narrated to me as feeling like an 'out-of-body' experience. It can involve a splitting of logic from emotion, of natural family from actual family situation. This is a mobility in and out of the self, and what one imagines one should be able to do. Whilst smoking a pack of cigarettes over successive cups of coffee, eating nuts and seeds in a three-hour interview conducted in my Jerusalem apartment, Dorit, a woman in her late forties told me about her experience:

Dorit: I was already 45 when I began. And the doctor at *kupat holim* [one of the nationalized health insurer's clinics] told me that my chances are very low, but they accepted me. I did two cycles of IUI [Intra-Uterine Insemination], which was completely natural, no pills. Then I did a few cycles of IUI with pills, to increase ovulation, and after a few treatment cycles the notion of IVF arose. They told me that at my age 'you do not do IVF', and that the chances are not better than IUI. But I was persistent and they accepted me to the IVF programme.

I once had great, excellent results and in the end it didn't succeed, so they stopped treatment. And I already started looking into the idea of egg donation, I did three cycles.

Michal: Here in Israel?

Dorit: No, no, I feel as if I'm telling someone else's story now [...]. Now, there is an issue here of [...] *it's a process of coming to accept that you are in treatment.* When I was doing the IUI I didn't, spiritually speaking, even want it to succeed. I had a problem with the sperm donation. I was sort of divided into two people, the logical person and the emotional person. The logical one decided to go for it because I didn't have any other choice. But the emotional person did not connect. It was as if I was two different people. And I felt sorry for myself I kept thinking [...] why do I [...] why couldn't I [...] why can't I [...]. Why, do I have to be different? Anybody gets married in their twenties, has children [...]. How did I get to this situation? (Dorit, interview, Israel, 6 March 2002).

Dorit was deeply frustrated with her situation. She was adamant that I understand the *processual* nature of coming to terms with having ova donation. That she wasn't 'this person' who would accept 'this thing' before she had gone through it many times. Despite accounts of the seeming out-of-control way in which Israelis undergo IVF, the women that I spoke to portrayed a painstaking decision process, as well as much social and familial pressure. She did not *feel* like a legitimate woman, mother or person because she would not be doing it in the 'normal', heteronormative way. She explained that she was also still, despite the decision to undergo egg donation, quite ambivalent about what her relationship would be to her 'future child'. Her frustration with what she frames as her own failure to succeed in the dream of heterosexual coupledom and childbearing and her own questioning of how she got to this situation, is indicative of some of the ways in which self-blame is integral to the experience of becoming an ova recipient. Many women expressed similar self-blame.

The affective aspects of taking an egg from another person and accepting the idea/prospect of this was often coupled with the knowledge that egg donation had come to a halt in Israel (due to the egg theft, see Chapter 6) and that the actual process of getting an egg would involve travel. Rinat, a 44-year-old soldier, referring to the fact that she 'needed' eggs and that there were none to be had in Israel at that time said: 'My whole world was destroyed.' The relief that she felt when she discovered the clinic performing 'reverse traffic' egg donation (see Chapter 3) was palpable in our interview. She commented:

> We did IVF and many different tests over a period of several years. I had hysteroscopy, biopsy. And they weren't able to produce any more eggs at some point. So October of last year they told me that I will require egg donation. The doctor told me that my only chance was to go sign up at all the private hospitals that provide egg donation. *And my whole world was destroyed.* I understood that there weren't any – any – donations happening *ba-Aretz*, so I didn't have a chance to get pregnant. And then I saw in the paper about two months ago, an article about egg donation. And it said they might make it illegal, so I contacted Ofra from the patient advocacy group CHEN (the patient support group). And she told me of this place which still performs ova donations. And I knew that I had other options like going to Cyprus but I already knew that there was this simple option where I didn't have to go back and forth to a country that is not mine and so that's how I got here to talk to Orit about the treatment procedure. So I did my first treatment cycle. And that month I started bleeding,[7] so they said they'd postpone the embryo transfer to the next month. Then the next month the uterus was ready, 8.5 mm, this was great. We did the transfer, I got pregnant on the first try and then between the sixth and seventh weeks I had a miscarriage. (Rinat, interview, Israel, 28 August 2002)

Beyond the reticence to travel abroad for egg donation (as opposed to travelling for holidays, which Israelis take abroad in very large numbers), which very many of my interviewees expressed, was a concern with the 'genetics' of their potential future child. This was one major disjuncture between lay understandings and hegemonic Israeli state policy at the time, which dictated that gestation of an embryo and foetus is what confers motherhood. Here, the egg recipients argued repeatedly that they were like an adoptive mother to their future child, or that they worried about the donors' genetic history affecting their future child. This is another instance in which the interviewees in 2002 anticipated the new Egg Donation Law of 2010. They often struggled with balancing the fact that they wanted a child with the fact that that child would not be related to them. Dorit expressed these ideas as follows:

> It's a process. At the beginning I was against it, but then I connected with it. I connected with the sperm donation. Once I got to the end of the IVF it was the child, any child that was important to me. Okay,

> having a child that is genetically mine is important, very important. But I knew that I wanted to have a child. So I started to make all the inquiries about what is needed for the egg donation. I was very much against it. But I had already put myself into a state where I wanted a child and I didn't want to give up this hope of having one. So I thought, 'Okay, you want to have a child, you cannot have a child of your own, you have one option, egg donation. If you don't want to have a child then go home, go have fun, do whatever! But if you want a child, this is your option. It's not the best option but it's a very good option. You will have a child and probably nobody will know it's not yours.' But I still didn't connect. I went through all the procedures but the emotional side of me really objected to this because I would never have a genetic child of my own. And I don't know why this bothered me so much. What's so fucking important, excuse me, about my own genes? Are my genes so great? I am a horrible person [laughs]. What's so great about my genes? There's no logical explanation for it. But it was crucial for me to accept that I would not have a genetic child. I mean every person wants his/her own genetic child. 'Your own flesh and blood,' so what? So it's not your own flesh and blood, so what [...] but I wanted to know why this was so important to me and I, I don't know why it is [...] and I still don't know. (Ibid.)

Dorit so movingly questioned her own need for a genetic connection with her future child. 'Natural motherhood' seems so deeply connected here to flesh and blood in the traditional Western kinship sense (Strathern, 1992a). And this flesh and blood would, according to rabbis and doctors, of course be 'hers' in that she would produce it in her own body, with her own cells if we took a strictly 'biological' view on the process. Yet, her 'local biology' is deeply genetic. Blood and flesh *are* genetic in Israel.

Another woman, Yudit (discussed further on) told me about her daughter, a 'ginger-haired' girl who resembles her father. Yudit contrasted this with her own brown complexion and dark-brown hair. Her daughter was born from a donor whom Yudit credits with having 'saved [her] life'. Like Rinat, above, her very existence is brought into question by the lack of a child. Indeed this becomes clear when we see that the Israeli phrase for a couple who does not have children is '*hashukhei yeladim*', which roughly can be translated as 'a darkness from lack of children'. When this daughter was born, the response of the parents of her husband was, 'Oh well, no matter.' Yudit stated that for Sephardim, the desire for a son to continue the name is very strong. Despite the family's reaction, *she* was happy with this daughter and wanted another child. The difficulty, she explains, is the shortage of eggs in the country. She had registered with a clinic in Tel Aviv. They import eggs from Romania. Yudit begins to tell me her thoughts on how she feels about the fact that the child will not be genetically related to her:

> That's something I'm always trying [...]. I look at my daughter, so many of the things in her character, that I try to understand where they came from, whether they came from my husband and [...] certainly from me she has nothing [...]. But there are things, in her behaviour, because they [children] try to imitate us and they [...] it's not because it's found in their genes. But, for example, she's a stubborn redhead. Now, my husband is really not stubborn, he's a very *vatteran*[8] redhead and I'm also *vatteranit*. So it interests me sometimes to think, 'Oh, maybe the *mother* is very stubborn' [my emphasis]. [...] There are things that I think I should have maybe checked beforehand, about inherited traits [...] but I presume they check these things when they're screening [donors] for inherited diseases. So I don't need to worry. But these are things I want to know. Maybe I can't meet the woman, but at least to know some things about her. (Yudit, interview, Israel, 5 March 2002).

Here, Yudit argues that genetics is not important, but inheritance of behaviour seems to be an idea for her of what gets 'stuck' from the contribution of the donor. Importantly, Yudit also calls the donor the 'mother'. So in some sense for Yudit, her child has two mothers. I ask her how she feels about ova donation at this stage. She states that since giving birth, the idea that her daughter was born of ova donation had been obscured, 'it's not something I need to talk about, we don't even mention it at home. Like, it's something natural, really.' For her it is natural to have a donor.

Yet something here perturbs me. On the one hand, it is important that the local biological isn't so strong that it sticks in the same way to each of the recipients that I interviewed. On the other hand, there is something here about what Yudit tells herself and what she tells me about the importance of a genetic connection. I cannot and do not wish to divine which is 'truer'. Both are.

Part of the issue is that the women whom I interviewed are embattled within themselves about the origins, geographical and genetic, of the eggs. The wording here is crucial, they are in a battle. So, more central here than genetics perhaps is the idea of the symbolic war involving destruction and survival deeply embedded in their 'can't have a baby' narratives.

There have been shootings and bombings daily in the settlements and at army posts along the roads. Many people have died. The government has been striking. I've become addicted to Israeli TV and this is not good! (Fieldnotes, 22 February 2002)

Waging a war to make a baby: militarizing egg donation

One day I was at the clinic in Tel Aviv observing and the recovery ward was full, about six beds were occupied. Sharon, the nurse who emigrated from the USA and with whom I sometimes chatted in English, ushered me in to observe from behind the secretaries' desk facing the people who were waiting. Orit came by at one point and commented that I could not really see anything from where I was sitting, so we found a convenient place for me to sit down. She and I chatted for a bit. Later I 'hung out' in the recovery ward. A woman named Shoshana was wheeled out of surgery, having just had 'embryo transfer'. The nurse, as she wheeled the patient out of surgery, told me that Shoshana was willing to converse with me. I therefore followed the nurse and patient into the cubicle where the woman was placed and conducted an interview. Shoshana was a 49-year-old woman at the time of her embryo transfer. She told me that she was a grandmother and had two daughters aged 26 and 29. She was divorced from her first husband in 1994 and had remarried since then. Her current husband also had children from a previous marriage. She explained that her treatment/pregnancy history was long. In her first round of treatments, she had become pregnant three times but the pregnancies did not advance. After having had an extended break from treatments she decided to try again. In this second round of treatment cycles she had ten IVF cycles and eight pregnancies. None of them developed to term. One pregnancy developed to 24 weeks, but she said that her 'cervix opened and the babies died'. They were twins, and '*ha-kol halakh*', 'all was lost'. She commented that the only person who knew that she was having egg donation was her husband. If people asked her what she was going to the clinic for, she would tell them that she was going for a new IVF treatment. She chose not to tell the rest of her family, 'because they will call [her] stupid'. Her co-workers did not know because she has quite a senior position and feared how her attempts at pregnancy through egg donation would appear. She told me that she was of 'Romanian extraction'. We talked about how poor Romania is today. She then told me that when her family arrived in Israel (in the 1940s) they lived in a *ma'abara*, a temporary camp for new immigrants.

I asked her how she came to seek treatment at this clinic. She told me that she used to be a patient of one of the doctors accused of stealing eggs. So she left that clinic and came here, *'ki lo hayta berira. Nisgar ha-berez,'* 'because there wasn't any other option. The tap was closed.' And so she was compelled to come here because she knew that it was the only place that was performing egg donation at the time. She had had three of four ova-donation cycles. This was her first try at this particular clinic. Shoshana commented that 'it was not easy to decide to do egg donation because [her] husband found it hard to accept'.

When I asked her about the genetics of the egg, she said that she was a lot less bothered about that, because 'the goal sanctifies [*mekaddesh*] the means'. And 'when you want something you overturn worlds'. But, she added that not knowing the origin of the egg was, in fact, preferable. She told me that, from a Jewish perspective, it is better to have an egg from a non-Jew or an Arab because then there's no chance of the child that is born later marrying its genetic sibling. Most of the Jewish women that I had interviewed prior to this (and indeed after this as well) had stated that they could never accept an egg from a Palestinian or Arab woman (see below for a discussion of the ambivalence surrounding this issue). Here, the 'means' of acquiring a child may be slightly problematic from her perspective, but the end goal makes it acceptable.

> So this morning there were armoured vehicles – army police and barricades at the hospital. Because of the *piggu'im* yesterday I took a taxi. What happened yesterday was that two young female soldiers were killed in Be-er Sheva. Kassam missiles were fired into Israel. The Israeli response today was to fire missiles into Gaza city, ruining buildings and wounding 25 people. Later, Dr B told me there was a rumour that there would be a *piggu'a* at the hospital, hence the military presence. On the news we saw that a quadruplet was born in Nablus today. Their names were Yasser, Arafat and the last two were named after martyrs. (Fieldnotes, 11 February 2002)

I returned to the issue of Shoshana's existing children and clarified how many she and her husband have between them. She answered, 'Three, but none together.' I asked why it was necessary to have one together, and she replied, 'Because it's a family, it's togetherness, it's a cell, nuclear.' We then return to the topics of treatment and the donor. She told me that she did not wish to know anything about the donor except her age. She waited six months before her eggs were ready (which contradicts Dr Shmuel's and the brochure of the clinic's representation of the treatments as 'fast'), and the treatment itself cost her NIS11,000, approximately £1500 (US$2337). She told me that she does not know where she got the courage to do this but that, *'Ani 'osa milhama le-havi yeled'*, 'I am waging a war to make a child.'

> A 'suicide bomber' was 'neutralised' in a café on Emek Refaim. 14 people were injured in the West Bank town of Ariel in a suicide bombing. This morning I woke to the sound of loud booms. (Fieldnotes, 7 March 2002)

I wondered after interviewing Shoshana what it meant at this particular point in time of the Al Aqsa Intifada, with almost daily suicide bombings, to say that one is waging a 'war to make a child'. The symbolic resonance was so strong, and seemed even more than the usual militarization of Israeli IVF and pregnancy documented by other anthropologists (Ivry, 1999). To make her relationship with her husband truly 'nuclear' she must wage war against the limitations of her body, her age and availability of reproductive materials.

For some egg recipients this created a militarized bond among themselves:

Dorit: It's uncertain whether the women who go together to the clinic abroad will have eggs from the same woman, but we joked once that our children will be siblings. One day the doctor rang her [another oocyte recipient she was with at the hotel] and I couldn't hear exactly what the conversation was but then he rang me and he said, 'I want to tell you that I'm not happy with the fact that the hotel put you two in adjacent rooms.' So I told him that it was fine, we actually asked for this. It turns out he told all three of us the same thing. Later when I met with him at the clinic he said: 'Listen, I need to maintain a discreet practice. If you connect with each other on your own, that's fine, I don't have a problem with it. I need to maintain discreetness.' I told him, you have been very discreet, but you should know, that this is a need. This is something that comes from underneath. It's something we don't have control over, an extant need. And I told him, 'Listen, I'm here [in Romania] a third time, and it happened all three times. And by the way the woman that was here with me the first time, and she was there for the second time, and this connecting up with other women had happened to her previously. It turns out, this happens to the women who come here on their own, whether they are single or married, because the husbands are [usually] not there. The 'problem' is the women's, even if it's really the man's sperm which is the real problem, it is the women who go through egg donation. And it is the woman who will never have her own baby. And it's a sorrow shared. You wanted Jewishness? 'A sorrow shared is a sorrow halved.'[9] So, I don't know if it is half the sorrow, but it helps. It's like, a hug, it's, it's [...] psychologically, you make of it what you will [...] I saw one of the women at the hospital recently, and how we hugged, and I hope we will be travelling together. It's not like you were with someone on a trip/holiday, it's not the same. Understand? It's not like you studied with someone and you have common memories, joint experiences – or, it's not quite the same. There is something not good here, very difficult, that connects between you. The pain, the knowing that you will never be able to have a child of your own. This is the glue. And I'm telling you, this happens on every trip.

Michal: Do you stay in contact when it ends?

Dorit: Sometimes you stay in touch a little. Me, less so. I personally have a problem, I don't know how to stay in touch, but they ring me, yes. The situation in which I met them is not a regular one, that a friendship developed in. There is something specific, very, very painful. Very, very intimate. For example, one of the women got pregnant and her entire family does not know it's an egg donation! I know it. I have a secret about her. I know something about her that no one else knows. And I'll keep my mouth shut about it and I won't even mention her by name [to you]. Everyone who saw her in Bucharest knows what she did there.[10] Her family doesn't even know she was in Bucharest. She's trying to get pregnant for seven years and she's pregnant now. That's it. Understand? Therefore there is something, again, it [...] we in Israel say: 'We parachuted together on the Mitleh' [...]. In the Sinai War, of '56, the [IDF] troops landed on Sinai. 'Mitleh' is the name of the place. During the operation, during all our wars, and *Barukh Ha-Shem* [Blessed be God's name], we had enough war, never except once, regular fighters parachuted for an operational purpose. Usually it is paratroopers, but not this time. The only

> time in Israeli history that regular fighters had to parachute into the battlefield. The youth of today probably don't know about this but it's like [...] we have something connecting us, it's something secret something very intimate, very discreet and something very, very painful. Awfully, awfully painful. And this connection is made, temporarily. (Dorit, interview, Israel, 6 March 2002)

What is close, what is too close? What must be kept apart, by whom and how? Egg donation and IVF have been theorized by Dorit as being about a deep pain, a kind of bonding of sisters, and a military bond. The phenomenological experience of being, becoming, not wanting or wanting to be an egg recipient tells another story of the relationship of some women to the Israeli state. The kinship between recipients is such a strong bond according to Dorit. It is almost national. Familial. But then it's also militarized.

> On Friday 12 April Tahani Assad 'Ali Fatuah a pregnant Palestinian woman, aged 35 who works as a pharmacist and lives in Nablus went into early labour. She was in her eighth month of pregnancy after undergoing fertility treatments having initially tried to conceive without assistance for four years. Nablus was under curfew on this day. The Red Crescent was called several times but was unable to reach her in time. A local doctor was called in to deliver the baby at home. The baby was born and then quickly died minutes after being born, having required an incubator. The hospital was two kilometres from her home. B'tselem attributed cause of death to the IDF who blocked access to medical assistance. (Summary of B'Tselem, 2002a)

Intimi **(Intimate)**

'Terumat beytsit ze davar intimi, hem lo yedabru'[11]

On a Friday morning in 1984, donors and recipients sat together in a waiting room at an Israeli hospital. They had answered an advertisement for ova donation that had appeared for the first time in the Israeli national newspapers *Yediot Aharonot* and *Ma'ariv*. All of the women, donors and recipients, were being treated with IVF. Even before the IVF law came into effect in 1987, it was agreed that egg donors should only be women who were undergoing IVF themselves. The oocyte extractions were performed by laparoscopy, using an external ultrasound monitor (see Chapters 2 and 3 for contemporary extraction procedures in Romania and Israel). By this time, sperm donation was being practised in Israel and around the world, with a great emphasis being put on donor anonymity (Landau, 1998). Here, however, the donors and recipients sat together in the clinic's waiting room. Ova donation was not yet shrouded in secrecy. Indeed, in this, the first case of Israeli egg donation, no protocols were in place in order to facilitate anonymity for donors from recipients. According to the social worker present at the time, this lack of anonymity was a big *fashla* (colloquial, a big failure, screwup). At its inception in Israel, there were no attempts to make ova donation secretive at an institutional level. Today, Israeli ova donation is practised with many levels of secrecy and is often narrated as 'intimate'. This occurs in everyday talk, clinical practice, responses in interviews and at the national legislative level. Israeli biopolitics means that ova donation is institutionalized as a secret that must be secured, and everyday discursive practices narrate it as an intimate thing that cannot be spoken about.

Before my arrival in Israel I was told that there would be no women for me to interview about experiences of ova donation because of the 'crisis' of an 'egg shortage' (Chapter 6). When I did arrive and spent several months observing IVF clinical practices and interviewing doctors and nurses in the Jerusalem fertility clinic, they told me repeatedly that, 'ova donation is an *intimate thing*, they won't talk to you about it'.[12] I found this idea quite strange at the time since it was clear to me that Israelis do talk about 'intimate' things in quite public ways. Already upon my arrival, and increasingly throughout the duration of my research, I watched countless television programmes about issues of reproduction, sex and the body. Family members whispered to me about so and so who 'must have had egg donation'.[13] Despite these warnings of the intimacy of egg donation (and that therefore it was not something that people spoke about), people *did* talk to me about this issue.

Given the importance of motherhood and 'the family' in Israel, it is no surprise that women who have experienced difficulty in bearing children attempt to have them with the assistance of technology.[14] It is further not surprising that there might be a certain level of 'stigma' attached to the 'inability' of women to have children. Yet it seems that Israeli ova donation is surrounded by discourses and practices that construct it as an intimate thing, but at the same time constantly reveal it. Can it be that the discursive construction of ova donation as 'intimate' is closely connected to the discourses that construct Israel as 'pronatalist'? Is the 'intimacy' of ova donation connected to the repetitive refrain of 'Israel has the most IVF clinics per capita in the world'? Is all this talk of '*intimiyut*' a way of making the nation and the border something that is both mentionable and unmentionable?

> Suheir Shhada was 17 years old, married and seven months pregnant. She lives in a village in the district of Ramallah in the OPT [Occupied Palestinian Territories]. In order to get to her local hospital to undergo routine medical tests she had to cross a checkpoint into Ramallah. On her way back she left the taxi, a couple of hundred metres from the checkpoint (which is as close as they stop) and walked toward the checkpoint. Along the way there was a tent and two army jeeps with IDF soldiers inside preventing 50–60 people who had been to Ramallah from returning to their villages. As the jeeps started approaching the people, the people began to run away in the direction of the villages. The soldiers began to fire and hit Suheir on the side of the head with a rubber bullet. (Summary of B'Tselem, 2002b)

Rina and Albert live in the economically deprived town of Ramle. I went to their house late one evening by intercity minibus, despite warnings that this was a dangerous place for a young woman to visit alone. We spoke for two hours in their home one evening. Rina, in her mid forties, was one of the few pregnant ova recipients whom I had met in Israel. As we talked, a friend of theirs and Rina's teenage son from a previous marriage, walked in and out of the room. It was clear that Rina and Albert did not keep their IVF treatments hidden from friends and family. But the egg donation was not something that they shared with these individuals. The boundaries of what was *intimi* to tell seemed to shift from the IVF *technique* to *knowledge* about receipt of body parts:

Rina: [A]bout the egg donation, we didn't tell them these details. It doesn't matter.

Albert: From my perspective, it's not a secret. But it's personal; you don't run and tell people.

Rina: I told my friends at work that I am in treatments, but not about the egg donation. It's not a big deal. Once it was a taboo but now lots of women do it. (Rina and Albert, interview, Israel, 16 July 2002).

Here, IVF is not considered too personal to tell. Yet regarding ova donation, their protestations that 'it's not secret' and 'it's not a big deal' (but they still won't tell, despite the fact that it is commonly done), seem to indicate that there is still something taboo about receiving eggs. For the majority of ova recipients this secrecy extends to their future child as well. For instance, Shoshana, who was still trying to become pregnant and whom I interviewed in the conference room of a private clinic said: 'We will not tell our child he is from a donated egg. Why should he know? This is not interesting. This is not an adopted child! It has no relevance.' The statement that it is 'not interesting', sitting alongside what many women told me regarding their concerns that the child would not be 'their own', can be read as a synecdochical moment between the inception of the state and their own conception. They would not read it as such, for certain. This is one small moment of resonance between the body and the nation.

Yudit has one child from ova donation and is trying for another. She works in a large hospital and identifies as descended from India. She told me, 'I'm sort of like a surrogate. It's not exactly mine.' The fact that she will carry an embryo to term and give birth to a child does not sit easily with her because there will be a missing connection with the child. She is not the genetic parent. Yet the phrase 'not exactly mine' indicates her ambivalence. She accepts that the child will at least 'partially' be hers. So her privileging of the ovum as conferring identity to the child is incomplete.

Dorit said, 'having a child through egg donation is like giving birth to your own adoption'. Such statements indicate the importance of genetic relatedness to Israelis. I have introduced them here for the purpose of linking them with statements about the 'irrelevance' of receiving donated ova. Clearly ova donation carries some relevance, otherwise it would *not* be kept secret. There is something else going on that may relate to the importance of having a link through the ovum from mother to child. I am referring here to the matrilineality of Jewishness, and the importance of Jewishness to the nation.

In Israel, the 'Public–Professional Committee on the Subject of Ova Donation' that deliberated the matter of egg donation in the year 2000 before tabling an Egg Donation Law was intent on having a secured donor offspring registry. This would be a database that contained the identities of donor offspring for the purpose of preventing future genetic sibling marriage. A company called 'InfoFORT' was hired to present the committee with the plans for a secured database.[15] Their presentation to the committee opened with an image of a blue 'ovum' secured by a golden lock (Figure 4.1). The document describing the registry lists its purpose and the ways in which it can be used. The cost of devising this registry was estimated at US$132,950, with an estimated extra

10 per cent for unforeseeable costs. The annual maintenance of this registry was to cost US$208,734 (Public–Professional Committee on the Matter of Egg Donation, 2000). The emphasis of the proposed database was on security in maintaining the secrecy of the donor offsprings' identity.

The very fact that such a registry was proposed is not surprising as there is secrecy and a desire to maintain information on gamete donation worldwide. Yet I want to suggest that guarding and securing the identity of the donor offspring (as 'illegitimate' children of their parents) materially semiotically stands in for the extensive Israeli practices of securing and guarding the nation (other, more literal, examples are the security 'fence' or rather, the wall being built to separate the West Bank from Israel).[16]

Similar to the case of egg donation, secrecy is also very strong in Israeli sperm donation. This secrecy in sperm and egg donation contrasts, somewhat to the case of Israeli adoption, where one might think that a similar level of secrecy would be present. Yet there is a certain level of openness regarding the passing on of information about adoption

Figure 4.1 Egg in lock
Source: InfoFORT (Public–Professional Committee on the Matter of Egg Donation, Israel Ministry of Health Online, 2000).

(adoptees and certain state institutions have access to a register) (Landau, 1998). It has been argued that secrecy in donor insemination is enforced by physicians as a strategy for maintaining 'professional power and male supremacy' since selection of donors is in the hands of physicians (Birenbaum-Carmeli, 1998). Landau argues that the secrecy is an attempt to maintain the 'traditional family with one mother and one father to prevent undesired input by the donor parent' (Landau, 1998: 3272). From an anthropological perspective, Birenbaum-Carmeli and Carmeli consider practices of secrecy in sperm donation as a set of ritual practices that anchor the myth of the family within Judaism to everyday life in Israel (Birenbaum-Carmeli and Carmeli, 2002). While a more extended discussion of the problematics of secrecy in gamete donation is important, the vantage point of this chapter follows the analysis of Birenbaum-Carmeli and Carmeli and the assertion of Landau regarding the maintenance of the 'traditional family'. In the rest of this chapter I extend these discussions towards a deeper interrogation of the biopolitical aspects of secrecy and *intimiyut* along the terrain of the border.

Like other forms of migration, such as travelling to seek work in wealthier economic zones, 'repro-migration' can signal that the travel has not simply been done for 'fun'. Indeed migration due to global inequalities is an important reference point for seeing who becomes a source of reproductive stuff and who travels. As seen in the previous chapter many of the women in Romania whom I interviewed expressed the desire to migrate elsewhere because life was difficult for them in Romania. From some studies of 'transnational migration' we get the idea that migrating also involves never completely leaving 'home' – emotional and cultural ties are preserved (Glick-Schiller et al., 1995: 48). The Israelis who elected Romania as a site for getting eggs from often did so because they were actually from Romania before they became Israelis (or their parents were). As well, the children born to Israeli women from cross-border ova donation must be adopted by their birth mother in order to gain Israeli citizenship. Finally, parents of children born of egg donation, adoption or surrogacy often think about the child's relationship to the 'genetic' mother and her homeland, and often have an intention to 'preserve' that imagined connection (Wade, 2007).

Repro-migration can refer to things as well as people, and is meant to highlight the very real connections between citizenship, migration and reproduction; instances when people or things are 'stuck' and when people do not or cannot travel. A sense of forced travel and movement make 'medical tourism' an untenable concept best left to the businesses offering global egg donation services (Inhorn, 2011; Inhorn and Patrizio, 2009; Matorras, 2005; Pennings, 2004; Storrow, 2005). 'Cross-border reproductive care' (CBRC) centres more on the carers than the recipients. Concomitantly, the term 'reproductive exile' is a very specific concept that works well for the Arab and Muslim exiles and refugees of recent imperialist wars of Britain and the USA. It would be disingenuous to call Israelis seeking to buy eggs from Romanian women 'exiles' at this stage in Israeli history, although perhaps after the Holocaust this would have been more relevant to the Jews. The most relevant term to this case might be Lawrence Cohen's notion of 'clinical homecoming' (2011) or Inhorn's 'reproductive return' (2011). I have condensed these into repro-migrations. But it's the 'going' as well as the 'coming' that is at stake here in making Israeli borders. 'Repro-migrations' then, attends to the manifold movements across symbolic–concrete (material–discursive) bodily national borders that occur in Israeli egg donation without simplifying who 'these women' are. They are neither global nor local in the narrow sense (Das, 2007, cited in Lambek, 2011). There are 'local biologies' here that are very much about 'migration', the border and the state or nation (Lock, 1995). That is, how to deal with the nuance and complexity of the concepts of 'Israeli women' and 'experience' of egg donation, without counter-posing 'local' to 'global' (as if the two were ever separate) – without eliding the universal in the anthropological search for 'the particular'. More specifically, what are some of the personal and financial costs of egg donation? What is allowed to pass into and what is kept out of the imagined 'Israeli body'? These are questions that engage the process of writing and genre as much as they are about gender, race and nation.

Part II: Israeli extraction – racialized borders, difference and mixtures

'Michal, John and I were wondering if you would donate some eggs to us. You know that I'm both Sephardi and Ashkenazi, just like you, so you would be just the right mix.' I was 26 years old and just completing my Master's degree when I was asked by a professor to donate my eggs to her. We sat down at a café across from each other and chatted over lunch. I had sensed something of this sort was coming. Just like her! At the time I was so flattered. I told her how touched I was that she'd thought to ask me, after all, I had seen her go through a couple of failed IVF rounds and it was heartbreaking to see someone want something so badly and be so disappointed. I told her my initial fears about the effects of the hormones, the large aspiration needle and having a genetic child out there in the world that was not really mine. She argued that in her experience the needle was not so bad and that they would, of course, compensate me financially. Specifically, they would pay for four years of PhD studies plus all of my expenses. This was money that at the time I desperately needed in order to begin doctoral studies. I took a few days to think about the request and offer carefully. Finally I rang her up and said that, no, I was sorry, but I could not do it. I just couldn't imagine how it would feel to have my 'child' out there in the world, I couldn't escape my suspicion that the procedure would harm my fertility. And, finally, I was worried about the complicated relationship that I would have to enter into with a high-powered academic that could affect my career. I was fortunate not to have to choose this option, although it probably would not have harmed me physically. I was deeply saddened to be unable to help her. What remains for me now and what was most intriguing and unexpected about this encounter was the idea of our common 'mixture'.

One day, I sat in Professor Barukh's office at a state-run IVF clinic while patients were consulting him (see Chapter 5 for more on this state-run clinic). He was berating a young couple for not taking their hormones properly. 'You have to tell them what to do and then ask them: what did I tell you to do? Write it down,' he said to me after the couple had left. I had been listening to these sorts of conversations for an hour now thinking about how familiar and familial the tone was. But I was also hoping for something else to do. A few minutes earlier Barukh had been speaking to someone on the phone saying, 'yes, it's fine to give her eggs to a Jewish woman'. My interest was piqued, but as the couple was still in the room, I could not ask him about his telephone conversation. I did not need to wait for long as Dr Ezra knocked on the door of Barukh's office and let himself in. I thought that it had just been Ezra on the phone. Ezra looked at me and said, 'So, you say you want to see egg donation? Come with me.'

He took me to the recovery ward to meet Maryam. Eager finally to see some egg donation related practice, I too was keen for her to donate. However, having myself declined to donate eggs in the past, part of me was conscious of also not wanting her to do so [...].

Maryam is a Palestinian citizen of Israel who comes from a trendy village just outside Jerusalem called Abu Gosh.[17] This is a village inhabited mostly by Palestinian Israelis but left-wing Jewish Israelis live there as well. Maryam is 27 years old, works at home and her husband owns a local shop. When I met her I felt slightly awkward as I was aware that she had just regained consciousness after surgery and must be in pain. Ezra rushed into the recovery ward and told her that she had produced 26 eggs. He explained to her that the normal number is 5, and he asked if she would be willing to 'donate a few in order to help another couple become pregnant'. Maryam asked whether this would reduce her own chances of becoming pregnant. Ezra replied that with the number of eggs that she had, giving away 5 would not harm her chances of getting pregnant. Maryam agreed but she wanted her husband's consent as well. Unfortunately we could not find her husband. He must have wandered off during the operation. I searched the various waiting rooms of the hospital asking for him by name, but to no avail.

I returned to Maryam and we talked for a while about being the same age and living in such different worlds. She asked about my life and I asked about hers. Among other things, she told me that she was 'so happy that the government pays for [her] treatment, because [she] just met another Palestinian woman who lives in the Occupied Territories who has to pay for her treatments'. Maryam indicated that the other woman was in a very precarious position. She explained that, due to the expensive nature of the treatment, the other woman's husband might leave her and find another woman 'who isn't so costly'.[18]

Maryam told me that she was just beginning her eighth IVF cycle. She said that she was hopeful that, 'this one will be successful'. Maryam reported that her previous seven cycles were undertaken at another hospital in Jerusalem but that she, 'changed [hospitals] in order to change her luck'. Maryam and I continued to talk for a while and then I left to see what was happening at the clinic. When I finally found my way through the mazes of the hospital wards, the nurses were shuffling through files to find the phone numbers of prospective egg recipients. They found Yudit's number.

Yudit is a 37-year-old Jewish woman who has been married for twelve years. She began tests for infertility after six months of marriage. There was a lot of pressure from her mother-in-law. Her husband, she says, comes from a family that is 'slightly primitive'.[19] She works at the hospital as a computer programmer and systems analyst. Having had one daughter already through ova donation, she had been waiting for another egg donation for over a year after several failed attempts. The nurses telephoned Yudit and exclaimed, 'We have eggs for you!' They were rejoicing because the previous week they had to telephone her to tell her that she had a '*beta shelili*' or 'negative beta' (that is, she did not become pregnant in the previous cycle). They were certain that this new information would make Yudit rejoice as well. They began making all the arrangements, telling her to find her husband so that he could come in as soon as possible to provide the sperm (eggs must be fertilized quickly in current practice, while sperm and embryos can be frozen). Finally, the nurse said, 'Oh and one more thing: the donor is Muslim, is this okay?' There was a pause. Disappointment registered on the nurse's face. Yudit had rejected the eggs; she did not want a baby from an 'Arab woman'.

Time was said to be running out and the nurses expressed this sense of urgency about the need to find a recipient. Maryam's husband had returned and together they agreed to donate the eggs. The nurses soon found a recipient. They called the head nurse at home. She remembered a Muslim woman who was waiting for a donation. They found that woman's file and telephoned her. Her husband arrived within the hour to hand over his sperm. The eggs were fertilized in the laboratory and incubated for three days. So were the eggs that Maryam kept for herself. Three days later both the recipient and donor underwent 'embryo transfer'. Two weeks later both women received their pregnancy test results, both were negative.

I interviewed Yudit a week later. She talked about always having wanted to experience a pregnancy. But initially she had had immense difficulty dealing with the fact that, 'it's my husband's sperm, and I know it's my husband's child but it's not my child. It's like, like I'll carry the child, I'm sort of like a surrogate mother, it's not exactly mine.' But she took heart that at least the child would be her husband's. She said that she would have preferred adopting but that her husband was against it, because he was worried about what his family would say.

Once Yudit resigned herself to egg donation, she registered with a few hospitals. I asked her what criteria she requested in the donor. She said that she did not know what to put down, except that she knew that she wanted someone Jewish. I asked 'why'. She replied that:

> It's something [...] I don't have an explanation, but I didn't want a non-Jew. And we're also religiously observant so there was a Halakhic [Jewish legal] problem with it. We are Sephardim,[20] and follow those rabbis' rulings. And the Sephardim have a problem with ova donation: the donor has to be single[21] or not Jewish. I, I had difficulty with this, like, with taking a donated egg from a non-Jew because I thought to myself, in Israel,[22] it could only be from a Muslim woman, so I was very scared. I didn't want it. This is something I can't explain [...]. I *really* didn't want this. Like, I wouldn't have minded if it was a tourist who came to Israel, and she was Christian, this really didn't

> bother me. I told them this at [another] hospital. But they don't treat tourists there. They have either Jewish or Muslim Arab donors. So I said I want it only from a Jewish woman. (Yudit, interview, Israel, 5 March 2002)

But by the time that Yudit discovered that her Sephardi rabbi would not permit the egg donation it was too late. She already had embryos. Her rabbi said that the child would be a *mamzer*[23] but the Ashkenazi rabbi said that it would be permissible and the child would not be a *mamzer*. She commented:

> That was on the day they were supposed to return the embryos [i.e. perform embryo transfer]. And it was very hard for me, very, very hard. I said, okay I'll do the 'return' but I really hope it won't 'take'. On the other hand, I really wanted to become pregnant. And I had three really hard weeks. I would get up in the morning and think, 'Oi, really, if I give birth are my children *mamzers*? Or, should I follow the Ashkenazi rabbis now?' But then the family would start asking 'What do you mean to say we are keeping *mitsvot*[24] according to the Ashkenazim? We are Sephardim at home!!' This was very difficult for me. (Ibid.)

Despite her reluctance about the pregnancy succeeding, she said that she took all the 'necessary' medication, including pills and injections, 'but prayed a lot for it not to "take"'. It did not 'take' and she subsequently underwent 19 cycles of IVF, after which the physician told her to stop, not to waste her time. Her response was to sign up at the Herzliya Medical Center (over an hour away from Jerusalem), a private clinic that is well known for its egg donation programme.[25] Always thinking that the following cycle would succeed, she continued for 24 cycles of IVF and explained, 'maybe for egotistical reasons, I don't know. But I wanted it a lot. Despite the fact that [...] [my husband and I] have a good relationship, truly good. It was still missing.'

Yudit felt that the treatment was very easy, explaining that, 'I took it all lightly. I didn't make an issue of, anything. Except for the negative outcomes, which were really hard [...] but, I didn't care, really, I didn't.' In fact, when the physician would send her away and tell her to take a break she would return and say, 'Listen, I'm frustrated, I want to start IVF again because I, I want to succeed in making something. And the *Kupat Holim*[26] won't approve it because there's a limit to the number of cycles they'll pay for.'

When I switch off the tape, Yudit tells me that she slightly regrets her decision to reject the egg from 'the Arab woman' but is sticking firm with this decision.

> I just walked into the city centre in Herzliya to deposit the rent money and stuff. I went to the post office to get a phone card, and on my way in I looked at the guard and looked away, so of course he was going to make a point of checking my bag [...]. So I said 'Shalom' and my accent always makes it clear that I'm Israeli and they have nothing to worry about. He says, 'Why did you have to come now and make me get up off my chair' and I retort, 'You could just check my bag while sitting down!' This is a friendly exchange and I think nothing more of it. Upon leaving the post office he calls to me and says, 'I want to ask you something.' I think, 'Here goes, you're nice to an Israeli guy and immediately he wants to ask you out for coffee.' But instead he asks 'What *'eda* [community] are you from?' So I ask him, 'What do you think?' To which he responds, 'If I knew I wouldn't have asked' (which is untrue because if he thought I was Ashkenazi I don't think he would have asked at all). So I say, 'I'm mixed, half Turkish, half Romanian.' (And I make a sign with my hands to say, 'Oh ho, so you see with what complications I have to deal with, both the "little Turk" and the "Romanian thief"' [...] both common stereotypes. And he responds, 'I knew there was something "little" in you [...].' A little Mizrahi he means.
>
> When you ask or indicate someone's extraction [or origin] *motsa* it is in order to categorise, mentally catalogue – label descent. This is a European practice of identity sense making – this is European kinship at its core!!!!! Israeli citizenship is based on kinship – you must be related to Jews in order to count as a legitimate citizen with real full rights. More so, you must be part of the Ashkenazi elite in order to count as a proper citizen with full realizable rights. So maybe it is slightly anachronistic to talk here of 'Jewish kinship' which is only one of the many cultural factors affecting relationality in and to the State of Israel. Maybe it makes more sense to talk about Israeli kinship – i.e. kinship practices and reckonings that make and uphold the state and its borders. Kinship has always been about race and fears of undesirable mixtures. (Fieldnotes, 21 June 2002)

Ethnographies of Israeli reproduction tend to parse Israeliness along lines of Jewish and non-Jewish, examining the role of religion in Israelis' thinking about making babies. Some have demonstrated an awareness of racism within Israel among Jews (notably, Teman, 2010; and Birnbaum-Carmeli and Carmeli, 2002). The tendency is to suggest that Israeli reproductive technologies are unique and particular. The question remains, when history, ethnicity and class are attended to, can anything more broad be extracted? I'm not searching for universalisms necessarily but more specifically trajectories of Israeli thinking on ethnicity and citizenship that are to do with the history of Ashkenazi Jewish settlement of the State of Israel, the necessity to bring in more Jews and to 'fix' those Jews, removing from them their 'barbaric' lesser Jewish habits and beliefs.

The focus on religion *rather than* ethnicity by other ethnographies of Israeli reproduction eclipses some important non-particularities about Israeli reproduction (such as racism, classism, how markets are deeply invested in it, war and so on). Israeli reproduction may not be as *particular or* 'local' as we might have been inclined to imagine when we saw Israel through the lens of religion alone. It is hoped that this book's intersectional approach widens that lens. The final conclusion of this chapter is that the focus on the particular tends to (perhaps unwittingly) fall in line with hegemonic Israeli nationalism: tropes of victimhood and survival leading to the need to be militarily dominant. This chapter shows that ethnicity is a highly relevant category to Israelis, and that it operates in similar ways to some Euro-American ideas because of historical ties to European migration and imperialist projects.

'Difference and Mixture' are egg donation imaginaries that make palpable: first, some of ways in which Mizrahim and Palestinian embody the borders (that is to say, the border is elsewhere than typically conceived); and, second, Eugenics and Europeanness (including Euro-American kinship thinking).

Mizrahim on the borders: unintelligibility and the creation of 'Others'

I had been observing for a few hours and doing some volunteer filing of patient test results at the UFC clinic (see Chapter 3). In the recovery ward there were only three patients but the rest of the beds had the curtains drawn, making it seem as though the ward was fuller than it actually was. When I entered the waiting room I observed two women sitting. I had passed them several times and noticed that they had been sitting there for a long time. I asked the secretaries, Gila and Neta, who the two women were and for what treatment they had come to the clinic. Gila's response was, 'The women are "Bedouins".' She said that she had 'no idea' why they were at the clinic. Neta then explained to me that that there was no way of communicating with the two women because 'these women don't speak Hebrew'. I wondered why the secretaries had not tried to speak to the women in English, but I did not ask them. I then mentioned to Gila and Neta that I would be interested in interviewing the women who were waiting. Gila and Neta said, 'Good luck' in an unhopeful tone of voice. I contemplated this exchange and thought about my next move. I was unaccustomed to approaching women on my own, as clinic staff usually approached women on my behalf as a matter of discretion. Then the two women rose and entered one of the physician's offices.

I asked the secretaries once again why the women were at the clinic and what treatment they were receiving. It seemed impossible to me that the secretaries would not know such important details. The only response that I received was that the women are 'Bedouins'. My confusion was compounded. The secretaries seemed to be refusing to answer my question. I was not accustomed to such secrecy by that stage of my research at the clinic. I was certain that the two women who had entered the physician's office must have been here to receive some sort of fertility treatment and that they had made an appointment with either Gila or Neta in order to be able to meet with the physician. The secretaries' inability to communicate with the women meant that the patients were outside the realm of communication. I turned to the manager for assistance.

I found Orit and asked her whether the 'Bedouin women' spoke Hebrew and whether she thought that they would be amenable to being interviewed. Orit's reply was, 'They're sort of Bedouin–Ethiopian.' By this time I was completely confused and extremely frustrated. I was not succeeding in getting an interview with the two women and furthermore I was becoming even more confused by the replies that I was receiving from the clinic staff. When the two women came out of the physician's office they were still talking to the doctor in hushed voices. When they finished talking to him they immediately left and I did not have a chance to approach them.

The staff's responses deflected my requests and instead they offered confusing labels for the two women's identity. I wondered, did they believe that the 'Bedouin' or 'Ethiopian' identifier would make me think that I did not need to interview them? Yet Orit's comment that the women were 'Bedouin–Ethiopians' was a conflation of non-Jewish Bedouins with Jewish Ethiopians. The clinic staff seemed to be conflating Bedouins, whom they view as not Israeli because they were not Jewish, with Ethiopians, who are viewed as the 'other' despite being Jewish.

To my 'diasporic' ears that were used to distinctions between Jews and non-Jews, Orit's conflation of Jews and non-Jews in 'Bedouin–Ethiopian' was particularly unusual. Yet her conflation seems to support Smadar Lavie's argument that Arab Jews or Mizrahim are marginalized along the lines of race and class (Lavie, 1996), in a similar way to the classification of Palestinians and Arabs in Israel and that Arab Jews and non-Jews in Israel 'are the border' (Lavie, 2011a). Indeed these findings also anticipate the legal encoding of the border in the Egg Donation Law eight years later. This failed attempt at interviewing two women in the clinic is a complex moment in which it is possible to think about how some bodies are seen as alien and illegible.[27] The borders of the nation, marking insiders and outsiders, were being created through the linguistic marking of the two women as 'other'. Or, psychosocially speaking, 'the "we" of the nation [is] affirmed through the difference [...] rather than against it' (Ahmed, 2000: 95). The 'Bedouin' women were constituted as 'strangers', with whom it was simply not possible to communicate. The women's 'difference' from the 'we' of the clinic and potentially the nation was created through their illegibility/unintelligibility to the staff. The women's dark brown skin tone was being remarked on as a semblance of difference from the majority of the people at the clinic at that moment, but the groups that they were imagined as part of were thought to be similarly different. In this case it did not matter whether they were Muslim Bedouins or Jewish Ethiopians. The border between Jew and non-Jew may after all not be so stark.

On another occasion I sat at the secretaries' desk with Gila and observed my surroundings. Gila, who had been the person who was most cold and suspicious of me at the start, had become very friendly. The desk was full of papers and in a state of organized clutter. There were two computers, two chairs, many papers and files, a fax machine, two credit card machines that seem to belong to two different companies and four telephones. After chatting with Gila, I left her to her work and went to observe the rest of the clinic. At one point Orit took me into Dr Glick's office, handed me a blank 'Patient Details' form and asked me if I would fill it out for the couple in the room. She instructed me to indicate hair colour, eye colour, height, date and everything else. 'Because', she tells me, 'the woman does not speak Hebrew very well. They're an Arab couple.'

It was with a mixture of relief and annoyance that I heard the request. I was finally passing as a legitimate person in the clinic – my academic status as a researcher was being accepted. But also, I was part of the unspoken European Jewish elite by virtue of being read as an academic.[28]

I peeked out of the physician's office at the couple. They were sitting in the waiting room. Before I had a chance to say 'no' to her, Orit placed the form in my hand and left the office. I wondered why Orit had given me this task. Was it to avoid the discomfort of asking the woman what colour her hair was under the headscarf that she wore (this seems like a strange prospect as many religiously observant Jewish women cover their hair)? Was it to test me to see if I would do what she asked? There seemed to be endless possibilities for why she had given me the form. I was confused by her request but I tried, nonetheless, to fill out the form, peeking out at the woman as I did so. I simply left blank the sections that I could not see or determine, such as hair colour and height.

While Orit's transferring the 'patient details' form to me involved in some sense my enrolment in the clinic (an ethnographer's dream), her reference to the fact that 'they do not speak Hebrew' indicates that while I was considered to be the 'right' kind of Israeli (Hebrew-speaking, well-educated, Jewish, middle class – read Ashkenazi), this couple was not perceived as such. What underlies this is the idea that the border is between Jew and non-Jew, when in everyday realities in Israel, that border is more blurred when it comes to non-dominant subjects.

'Irbuv shone ('a different mixture'): vitality, viability and the national body

> [T]hese children are amazing, every one, amazing, beautiful and full of *'hiyuniyut'* (vitality), curious and full of life, simply amazing [...] much more beautiful than the parents. (Orit, co-founder, Universal Fertility Care, Israel, participant-observation at the clinic, 2002)

When women first register at the private IVF clinic, along with many other forms, they receive a form titled 'Preferred External Features'. The women and men are told that this form is meant to provide the clinicians with a guideline on how to proceed when matching donors' eggs with recipients. Many women told me that they did not care about this form, they filled it out along with other forms provided to them and it was not something that they thought about a lot. Nevertheless, it is a significant document because all interviewees filled it out. Furthermore, it is an important artefact because the clinic insisted that all clients complete it. It seemed to be a technique for providing clients with consumer choice in their donors (see Chapter 3).

The 'Preferred External Features' form had seven categories: date, name of patient, skin colour, height, hair colour, eye colour and comments. The date, name and comments fields were open. The rest of the fields had several choices, with little boxes to be ticked. Under skin colour there were the options of: light, swarthy, dark and very dark; height: 1.6, 1.7, 1.8 metres; hair colour: blonde, brown, black, red; and under eyes: blue/green, brown, dark. It may be that the traits that women select are of those of a donor who is an idealized image of themselves, has a resemblance to their partner or who represents an idealized, imagined partner.

It may be the case that the selections that women make indicate a desire for something other than 'just' a baby. The selections that they made and the comments that they wrote on the forms were extensive and often indicated an internalization of anti-Semitic stereotypes about Jews. They rejected traits imagined to be 'too Jewish' (see Gilman, 1991). The traits that people rejected seemed to represent ideas about racial weakness, similar to those antiquated stereotypes of Jews found in late nineteenth- and early twentieth-century Europe. Given that many women wrote comments about these attributes and pointed out such preferences to me in interviews indicates that these selections are an important aspect of Israeli egg donation. These kinds of rejections indicate some of the ways in which racial difference and undesirability are performed in Israel (Birenbaum-Carmeli and Carmeli, 2002; Lavie, 1996; Motzafi-Haller, 2004; Shohat, 1989; Weiss, 1994, 2002).

'Positive eugenics': European bodies, European beauty

Nearly all of the women whose forms I examined ticked off the box for 'light skin'. Many added additional comments to emphasize this. Several women underlined the word 'light'. Some indicated that 'it should be white, but not a "strong white"'. Here there is an ambiguity as to whether this was a reference to skin, egg, donor or all three. The focus on white skin but 'not too strong' seems to indicate a concern that the child should not look too Middle Eastern and conversely not too 'un-Middle Eastern'. Several women placed an exclamation mark after 'light skinned' or 'desired light'. I read some of these comments as indicating a wish for a particular kind of normative Israeliness. Birnbaum-Carmeli and Carmeli (2002) have shown that preference for lighter skin is also found in Israeli sperm donation. Teman (2003), in her study of Israeli surrogacy, has indicated that an Ashkenazi commissioning mother whose surrogate was Yemeni, for example, expressed a wish that their child not be born with dark skin (although these women admitted that this was not 'really' a logical possibility). Furthermore, many of my research participants expressed their wish not to have 'black' babies.

In addition to the comments about light skin, many women also indicated that they did not want a 'large nose'. The 'comments' sections were riddled with remarks such as 'small nose', 'Nose: not big', 'without a vulture-like nose and without thin lips'. The repetition and vividness of these comments could indicate reactions to and internalization and reproduction of racist perspectives of the early twentieth century regarding 'the Jewish nose' (Gilman, 1991). In Israeli writer Oz Almog's account of the image of the '*Sabra*'(for a discussion of this term, see Chapter 6 here), there is evidence of such a reaction dating back to the early days of the inception of the state. Almog comments:

> The Diaspora Jew was described as the diametric opposite of the pioneer of the Land of Israel and was portrayed along the lines (and to a large extent under the inspiration) of anti-Semitic stereotypes. (Almog, 2000: 77)

This undesirable 'old Jew', Almog argues, was the 'Other' against which the 'new Jew' (or what he calls 'gentile Jew') modelled himself in the early days of Israel's inception. The comments that women wrote regarding not wanting the traits of a large nose and thin lips indicate that Israeli notions of beauty are predominantly associated with small noses, full lips and light skin. This resonates with contemporary Western hegemonic ideals of feminine beauty that have dominated Western society since the eighteenth century. As Almog argues, these are beauty norms associated with 'European Christian' ideals (Almog, 2000: 79).

For example, one woman, Shoshana, an Orthodox Jew, indicated that for the donor to be 'appropriate' she would have to have a 'normal IQ and be pretty and healthy'. Shoshana commented that:

> We were told that the donors are students, which already demonstrates a normal IQ. I hope we can build character by raising a child, even if, God forbid, the child has characteristics of hatred or jealousy you can change them. But what we did worry about was her appearance. Because, *Barukh Ha-Shem*, 'thank God', my husband and I are pretty okay-looking. And we expected [...]. I asked for [...] in our family we

> have all the shades: blonde, blue, green, I have no problem with this. So they should match me up with something really *mat'im*, [appropriate]. My fear is always that it would be something not matching. This is a very scary point. Like, my husband has heard of a situation where a black child came out of a white woman. Do you know what a *fashla* this is? So this is the worst scenario. But here in the appearance, we're not looking for it to be a beauty king, we're looking at least, for a nice appearance that you can live with. This is very important, because luck is from God. At least it should be appropriate.
>
> If you investigate a secular woman it would be interesting to see how you compare her issues [...] how she sees it from a secular point of view to how a religious person sees it, it really interests me [...] maybe she doesn't care, it's only important for her to bring a child into the world and it doesn't matter to her who, what or how, it's also possible that this would be the case [...] but we're for, if you bring a soul into the world then it should be a soul that will be what's good for us [...]. Let's say we had to have a child and the child was, God forbid, had a birth defect or something was wrong with it like these autistics, it's not easy [...] so we pray to God, what's good for us, give us. (Shoshana, interview, Israel, 6 August 2002)

Shoshana's comment about having a child that 'matches', suggests that she wants a child who will look similar to her and her husband. But the notion of appropriateness resonates with ideas about what counts as 'good enough', for her or for the nation. A black or disabled child would be considered very unsuitable.[29]

Dorit, an Ashkenazi woman, was more explicit about her desire for whiteness, and indicated that Israel is a 'racist' society and that for this reason they wanted a light-skinned donor.

Dorit: What matters is that she [the donor] should be my own colour. That matters.

Michal: The same colour of hair?

Dorit: No, skin colour. It matters, and I tell you why. This is also in adoption. I don't want a coloured child. [...]. And [...] I am not a racist, not at all. I love [...] I would like to have a Brazilian husband, a black husband or whatever, I would like to. [...]. Now, about the egg donation, it must be the colour of me because I don't know yet if I am going to tell the child that it's egg donation or not. It's something that I cannot answer yet. For now this is my own child, so it must have my own colour. [...]. A

> single woman, *had-horit* [single parent], the child initially faces a lot of problems, being a single mother child, and children are very, very bad. *Akhzari* [cruel]. I want this child, whether this is an egg donation or an adoption one, to have as minimum obstacles in his life as possible, because initially it's gonna have obstacles in his life. So I don't want children to [...] to [...]. If he is 'coloured', they are gonna tell me that he is adopted or whatever. They gonna give him a hard time. You follow my [...]. And I wouldn't mind. I don't have anything against Vietnamese or Philippines, Brazilians. Not whatsoever, especially not Brazilians, because I am crazy about Brazil. I've been to Brazil, I am crazy about Brazil, it is one of my favourite places on earth. And I like Brazilians, they are good looking, they are gorgeous. I would not like in my position to have a mulatta [*sic*]. Because of the problem of colour. Children are cruel to those who are out of the ordinary.[30] And a 'coloured child' to a single mother, that means he will have problems for his entire childhood. And I don't want that. A white child does not draw people's attention that much. A child that is 'different' because he is the child of a single parent is already going to face problems, so why add to that? The doctor told me he does not accept ginger-haired donors. He is concerned with matching donors to recipients' appearance. Now, the Romanian women are very beautiful. They are European in every way. I don't have a problem with them as donors because I have overcome the pain of deciding to accept donation and now I don't have any other choice. (Dorit, interview, Israel, 6 March 2002)

Negative Eugenics

Some recipients expressed concerns about the donor's medical history, as well as ideas about illness and 'contamination' of ova from HIV. For example, Rinat told me that, rather than what the donor looks like, 'the thing that was most important was that I would know that I wouldn't get a contaminated egg [with HIV], and that the embryo would not be contaminated'. Another woman, Esther, told me that she did not care about the appearance of the donor. Rather, for her, the health of the donor mattered most. She said that she worries about:

> [W]ho the donor is. It's clear that she's doing it for money, and that's legitimate. But who knows what her medical history is, and

> her family history. I don't know anything. That is, personal details didn't interest me whatsoever, neither intelligence nor looks. But I was interested in the medical and family backgrounds. I was afraid of this. That's why I avoided egg donation and continued doing the IVF cycles [even though she knew that they would likely 'need' egg donation]. (Esther, interview, Israel, 3 September 2002)

For Esther, it is the donor's medical history and family background that matters. Esther imagines the donor's history as being contained in the eggs that she is giving.

In the account above, of what traits recipients want and do not want, getting the 'right mix' seems to involve European notions of beauty, an ambivalence about the borders of the nation (what is allowed in and what must stay out) and ideas about health and intelligence.

For almost all of the women and men whom I interviewed the overriding concern was to have a baby. Yet the many competing logics that inform their negotiations of the process explode and implode geographies. The bridging of distances between these places being 'enabled' by physicians, laws, technologies and economic arrangements as well as imaginaries of the self and the 'other' despite widely publicized claims of free and easily accessible IVF, some pay US$3000 for eggs and others (Romanian donors) receive US$200 in compensation. The descriptions of recipients' desires regarding donor traits in transnationally trafficked eggs congeal aspects of the racialized national body. They do so in a way that incorporates ideas about hybridity and 'self-improvement' through 'mixing'. The parents of the ova donation children hope that their children will be a 'better mixture' than they are, and that they come from 'better stock', that is, European stock. Furthermore, recipients hope that their ova/children will not contain any undesirable elements. One day as we chatted, Orit told me, 'these children that are born from egg donation at our clinic' (that is from the combination of an egg from a Romanian woman and sperm generally from a Jewish Israeli man), are '*irbuv shone*', 'a different mixture'. She continued to explain that these children are, 'amazing, everyone, amazing, beautiful and full of *hiyuniyut* [which she translated as 'viability', although the word actually translates as 'vitality'], curious and full of life, simply amazing [...]. Much more beautiful than the parents.' These comments, that the children have 'viability', that they are 'different' and a 'mixture' draw on Israeli racial and nationalist thinking. I argue that these comments, considered in conjunction with the views considered above, indicate that egg donation and the relationalities that emerge are saturated with ideas about difference and mixture, which are key features of Western racial thinking.

The notion of viability, which Orit invokes, involves the idea of robustness and strengthening. Orit suggests that improvement will be obtained through mixture. As Haraway (1997: 233) has pointed out, racial mixing (at least in the USA) is both feared and desired; it involves elements of purity *and* contamination. Indeed the valorization of mixture is a feature of multiculturalism in many national settings such as the UK (Ahmed, 2000; Fortier, 2001). Ahmed (2000) argues that the national self is affirmed through discourses of accepting the 'other'. I illustrated earlier some of the ways in which this ambivalence regarding mixture, is present in ova recipients' perspectives on receiving ova. Here, I turn to look at how such notions of 'mixture' and 'difference' are central to Israeliness.

This idea of a 'different mixture' is an 'emic' (or local) term, which can prove useful in explicating the related processes of imagining nation, race and kinship. In positing the children as a 'different mixture', Orit may be fetishizing the difference supposedly embodied by those children. At the same time she is essentializing the genetic contributions of Romanian woman as bearing the 'truth' about the potential future children. Her comment implies that the parents require improvement. The ova recipients are imagined as 'lacking' something in terms of Weiss's formulation of the 'chosen body': a healthy, strong, militarized and reproductive body. The ova recipients can thus be seen as 'unchosen'. That is, Orit constructs them as unchosen. Indeed, clinic staff and recipients themselves occasionally referred to recipients as 'second class' or as lacking something.

The desirability of the baby comes from the imagined sense that it is a mixture of 'there' and 'here'; a different mixture and a mixture denying difference, embracing instead Europeanness. The geography of desirability must not be forgotten. It is not merely a global ideal of beauty that recipients were selecting. It was a distinctly Euro-American one. One woman indicated that she did not want someone from 'Gypsy extraction' but that the donor should be beautiful, with a slim body. Another said, 'not fat, not Muslim, preference for musical intelligence'. Several others indicated, 'not black or Ethiopian'. All of these statements suggest that the 'right' kind of traits do not include imagined 'others'. In part, the preferred external features form is a technology of racism, which extracts recipients' ideas about what is acceptable and not acceptable in the Jewish Israeli body. Based on the comments written on these forms, I started to ask recipients about whether they would accept eggs from Palestinian women.

Arab eggs

The selections of traits of their child, donor, egg, are also about border management. These semi-porous membranes allow ova from non-Jewish Romanians, but most often not from Arab or Palestinian women. Moreover, most recipients who wanted children with darker skin, hair and eyes, did not write any additional comments on the forms, not needing to expand on their selection. The people who do not choose Europeanness are not doing the same kind of national imagining. Being closer to 'the border' of the limit of Israeliness and to the literal borders some of the Mizrahi women seemed ambivalent about whether or not they would accept an egg from a Palestinian woman. I do not wish to claim that this is a great show of solidarity or a major sign of resistance. Successive Israeli governments have talked about the border as a space of anxiety for Israelis, constructing military and solid walls to protect the people and prevent anxiety. Some Israelis have come to 'embody' the border, having been settled there as immigrants and remaining as borderline citizen/subjects in the minds of dominant, Ashkenazi Israelis. These very women have been on the frontline of fire, guarding the borders as they do (symbolically and physically). The contradiction between what they feel and what they think that they ought to feel (for the nation, for their family, for their mother tongue, Arabic) is potent. Opening up these spaces of ambivalence indicates that although egg donation 'choices' are haunted by the past of European Jews, Arabness presents a more complex ontological choreography (Thompson, 2005) than I had previously imagined (Nahman, 2006). At the same time as Europeanness is being 'reproduced', Israeli women are making and unmaking the borders of Jew versus Arab. These moments of ambivalence are moments of hope.

I began by asking Ofra, the head of CHEN, the patient advocacy group, whether Jewish Israeli women have difficulty accepting eggs from Palestinians. In addition to indicating their feelings about Arabs, I thought that this question might also point to their beliefs about the importance of ova, given that Jewish law dictates that the gestation of the baby is what counts, rather than the identity of the donor. Ofra told me that some Orthodox Jewish Israelis were going to the Palestinian city of Nablus for ova. So, in some circumstances, 'Arab eggs' are acceptable to Jewish Israeli women. Yet when I asked ova recipients and doctors about this, no one could provide me with concrete details about how it was done. One nurse whom I questioned claimed that some Israeli women would accept an egg from an Arab woman, as long as she was from the educated classes. However, when asked if they would accept eggs from a Palestinian woman, ova recipients almost all answered 'no'.

Clearly women have varying ideas regarding their relationship to donated ova and prospective children. However, Ofra's suggestion that Jewish Israeli women *did* at one point go to Nablus for eggs is a sign that perhaps, when the situation is not as volatile between Israel and Palestine, then Jewish women may secretly accept ova from Palestinian women. Most of the women whom I spoke to denied the idea that they would ever accept eggs from an 'Arab' woman. Rina and Albert, who self-identified as Moroccans, had a disagreement about this issue during our interview:

Rina:	Regarding the religion of the donor, Albert and I didn't talk about it, so I guess it's not relevant. Of course it is important that she shouldn't be an Arab, *ya'ani*.'[31]
Albert:	*Davka* [especially] not an Arab? Why not? What's wrong with the Arab women?
Rina:	I don't know [...].
Albert:	To me it doesn't matter. [...] I didn't really think about it. I guess the doctor knows not to mix. It's like they're not going to bring me a black donor [...]. But we haven't thought about this much. [...] I know it's not an Arab donor.
Rina:	I wouldn't accept a donation from an Arab woman. I wouldn't take it.
Albert (to her):	If they told us, 'Listen, there's an Arab donor, take it or leave it,' I would say, 'Take it.'
Rina:	No, no way.
Albert (jokingly):	I wouldn't ask you. (Rina and Albert, interview, Israel, 16 July 2002)

The ambivalence among Jewish Israelis regarding Arabs may not only relate to assumed religious differences between Jews and non-Jews, but also to the myriad ways in which Arabness is assumed and rejected by Arab Jews. This is especially the case if one takes a feminist postcolonial perspective that the history of Jewish immigration from North African and Arab countries to Israel had an 'unsettling' effect on understandings of what Jewish 'culture' in Israel was or should be. With respect to reproductive practices, for example, Mizrahi Jewish women were judged negatively for not acting 'responsibly' and taking contraception to avoid unwanted pregnancies (Amir and Benjamin, 1997). There was and is a constant back-and-forth policing of what is appropriate around Arabness.

Some women whom I interviewed came to reconsider their initial negative reaction to the idea of receiving ova from Arab women. After I turned off the recording equipment, Yudit, who earlier rejected eggs from Maryam (who self-identifies as Palestinian) told me that she regretted this but felt that she had to do it. There was a kind of guilt about this decision. Rinat, another recipient (who did not self-identify in ethnic terms but as a soldier) articulates why she would not want an Arab donor. She stated that:

> [I]f the donor was from an Arab country this would be difficult because it is an enemy state, not specifically because it is an Arab state. This doesn't matter to me. I once had a Christian Cypriot friend. This doesn't matter to me and I wouldn't care if it was a Muslim donor. It's just [...] I don't know, it's in my head [...] you know what? Maybe I wouldn't mind [...] and what if there was an Israeli Arab couple in Israel who needed a donor, wouldn't they take from a Jewish woman? (Rinat, interview, Israel, 28 August 2002)

In the conference room of the private IVF clinic, Sharona spoke to me about being an egg recipient. She self-identified as Kurdish but said that she sounds Moroccan because she lives with Moroccans. Sharona cried a lot during our interview, something that she attributed to the pain of wanting a child. In terms of the donor, she said that she wanted someone who looked like herself. She did not care where the egg comes from as long as there will be a baby. It did not matter to her whether the donor was tall, dark-eyed or blue-eyed. But when asked if she'd mind an egg from an Arab woman, she said 'No, that no.' She did not mind if the donor was not a Jew because the act of feeding and raising the child to be Jewish would make that child Jewish.

Dana was a 48-year-old woman who has been married since she was 37 and has been trying to have a child since then. She found her treatments to be very costly and repeatedly told me that Maccabee, her health service provider, did not reimburse the costs. She self-identified as being Mizrahi from Afghanistan but insisted that this was not an important factor to her egg donation experience. Dana has had several rounds of egg donation, both in Israel through the private clinic in which I conducted my research and in Romania with another doctor. When we spoke it was in a very noisy recovery ward of the private IVF clinic south of Tel Aviv. Her experience of egg donation indicates that she believed that doctors were lying to her about the process, providing treatments that were unnecessary and sometimes not inserting embryos into her uterus despite telling her that they were doing so. Her trip to Romania felt to her like a waste of time. While we chatted she was negotiating the presence of a *mashgiha,* as she was religiously observant and wanted to adhere to Halakhic practice of IVF. When I asked her about whether she cared at all about the genetics of the donor (since other women and the clinic had found it important it became something that I routinely asked egg recipients), she seemed perplexed and stated that it does not matter at all, indeed she had never thought about it (Dana, interview, Israel, 26 August 2002).

Discussion

The practices of choosing and selecting, even if imaginary, paint a picture of Israeliness and borders that is central to both how the state functions and the relations of (re)production within Israeli society. Global topographies of race and economic inequalities scream out at us from these choices. Centrally here, is the racialized subaltern speaking subject and the political economic nature of egg donation.

Although Israel is generally seen as 'mixed', and indeed valorized as such by Zionists, it is commonly referred to as 'ethnically mixed' (as if 'ethnic' is the social identity mapped onto a biological 'race'). When I presented an early draft of this chapter at a major anthropology conference in Britain, a professor who also happens to be Jewish argued against my assertion that the boundaries of 'race' were being constructed by Orit. This professor suggested that in Israel people talk about ''*edot*' or ethnic groups and not 'races'. This professor was correct. People in Israel speak about ethnic groups. However, they do so in a way that reifies those groups and marginalizes Mizrahim in a way that indicates that ethnicity is a euphemism for 'race'. The disagreement between this professor and myself illustrates the contentious nature of identity politics among Jews and of the ways in which the role of racial thinking in Israel among Jews is effaced. Even to speak about racism, to 'see' race, results in being told that you have a 'chip' on your shoulder or are being racist yourself, reminiscent of how racism works in other Western neoliberal societies.[32]

Speaking of personhood and relationality in Israel is impossible without thinking of both 'race' *and* 'religion'. The many experiences, identifications and subject positions of Israelis – Jewish, Christian, Muslim, Palestinian, Ashkenazi, Mizrahi, migrant workers from Africa and Asia, new immigrants from Eastern Europe (this list is not exhaustive) – must be taken into account. Although not static, or reified, these different labels relate to different material and symbolic positionings. They are part of Israeli national imaginaries that form the 'mixed up' politics of race in Israel.

But to speak in this manner in Israel or about Israel is difficult. Israeli anthropologist Smadar Lavie has been one of the brave ones to stand out as an academic and activist stating the troubling facts of Israeli racism despite enormous costs to herself and her family (Lavie, 1996, 2011a, 2011b). She has argued that normative Israeliness is Ashkenazi. Jews who identify or are identified with descent from Europe are the Israelis who have privileged status within structures of class, labour and geography. She argues for coalitions among non-dominant Israelis (Lavie, 1996) whose cultures have been alternately devalued and fetishized. These economically underprivileged groups can be broadly identified as Mizrahim and Palestinian citizens of Israel. The typical split of Arab versus Jew is subverted in the idiom of Arab Jew with which some Israelis are now identifying.[33] She and other Mizrahi feminists (cited in Motzafi-Haller, 2004) have argued that there exists a kind of racism in Israel that lies *not* along the Palestinian/Jewish divide, but along the Euro-American/Middle Eastern divide. Lavie argues that while the non-European Israeli (or what Ella Shohat calls 'Third World' Israelis) is seen as hybrid, the Sabra (Ashkenazi) is seen as pure (Almog, 2000).

Considering how people are 'raced and erased' (Lavie, 1996) significantly undermines the view of Israeli politics as being only about conflict with Palestinians or among 'secular and religious' as if those divides were obvious and without a history (Asad, 2003; Mahmood, 2005). Lavie also critiques analyses of Israel that focus solely on Ashkenazim or only on Mizrahim as if they were a 'special case'. Instead she offers a hopeful strategy for transcending the marginalizing effects of the dominance of Ashkenazi Jews in Israeli society, recognizing that Mizrahim and Palestinians are marginalized along racial, geographic and class lines. She explains that all of these boundaries serve the interests of Ashkenazi Jews and the Zionist state. Lavie envisions (using an analysis of 'Third World' Israeli writers) a new version of Israeliness that takes into account the intersectionality of race, class and geography.[34] In this vein, it is crucial to pay attention to everyday idioms and images that evoke hybridity, mixture, authenticity, colonialism and history, as these mutually reinforce both Israeliness and Jewishness.

Israeli economic networks are built on kinship or family connections that facilitate transnational business connections (Nitzan and Bichler, 2002). The notion of *mizug galuyot*, or 'the blending or integration of the exiles', is a central element of Israeli nationalism and national policy (formally encoded in the Israeli Law of Return[35]). *Mizug galuyot* is a romanticized ideal of the mixing of Jews in the State of Israel. This romance around blending and mixture finds its way into everyday Israeli and Jewish culture through festivals, education and songs. For example, one song says, 'I was born to the tunes and songs of all the nations,' which refers to all the nations that form the mixture of Israeli society.[36] Hence, the idea of 'blending' and mixture seems to be a key metaphor of Israeliness. However, as in many other multicultural societies (such as Canada, Australia, the UK and the USA), the questions remain – what *kind* of mixture is desirable? Or as Anne-Marie Fortier has asked, 'who are the legitimate multicultural subjects entitled to belong to the national community and to speak in its name?' (Fortier, 2005: 561).

Historically in Israel, European immigrants have been seen as most desirable by the predominantly Ashkenazi establishment (Almog, 2000; Dominguez, 1989; Lavie, 1996). In addition, the Mizrahim ('Orientals') and more recently people from the former Soviet Union have come to be considered desirable immigrants by the state, who would raise the number of Jews in Israel and help Jews to 'win' the demographic war (Portugese, 1998) against the Palestinians. Yet in Israeli society, such immigrants tend to be marginalized in service jobs or lower-ranking labour. In addition, such marginalized immigrants are fetishized for their 'exotic' food and dances.

In effect, such immigrants do not comprise the 'right mixture' for Israeliness. They were constructed as primitive, illiterate, uncultured and generally dangerous to the Israeli collectivity (Dominguez, 1989). Russians and some Eastern Europeans are often criminalized and viewed as prostitutes. However, some Jews from Arab countries were considered to be desirable immigrants because they were perceived as 'more authentic' than Ashkenazi Jews. This implies their racialization as being more 'primitive' and having therefore retained more of the 'older' and 'purer' Jewish practices. Concurrently, they are mocked for having too guttural a Hebrew accent (Almog, 2000). Consequently, many 'civilizing processes' such as teaching 'proper' Ashkenazi pronunciations, were implemented by the Israeli state in order to create more acceptable citizens. These practices of 'normalizing' Mizrahim so that they sound more Ashkenazi, illustrate how hierarchies around race persist in Israel.

Expectations about 'the blending of exiles' rely heavily on the conflicting requirements of 'quantity' (more Jews) and 'quality' (the 'right kind' of Jewish population). More and more immigrants have been encouraged to come to Israel as part of the 'demographic race' (Portugese, 1998) against the Palestinians. Programmes were instituted amongst the 'Jewish Diaspora' to encourage young Jews to immigrate.[37] Yet consideration of the 'quality' of immigrants is a crucial aspect of this nation-building process. Immigrants who can prove beyond a doubt that they have 'Jewish blood' are considered likely to adapt and are therefore seen as acceptable (Dominguez, 1989).[38] Thus Orit's comments about difference and mixture are kinds of imaginaries that seem to resonate with certain national ideals.

Whilst national borders of the state are imagined as a line between Israel and the West Bank, for example, in reality this border spreads weblike throughout the Occupied Territories. Similarly in terms of Israeli citizenship, whilst one might imagine oneself to be an equal citizen, this might in practice be a much more complex issue – and in this respect at least, Israel is not really that different from most other Euro-American societies. This has led some Israeli feminists to interrogate the extent to which simply being Jewish really does ensure equal citizenship. Indeed the politics of Israeli multiculturalism and race unsettle the simple dichotomy of Jewish and non-Jewish in Israel. A perspective that takes into account difference, identification and exclusion contributes to my overall argument of the Euro-Americanness of Israeli society.

One important aspect of contemporary Israeli life that often gets either missed or completely reified is the issue of the politics of *difference* among Jewish Israelis. The history of migration of Jews to Israel from around the world has become a highly romanticized and politicized aspect of Israeli nationalism. Codified in law, in the form of the Law of Return, is the legal right of Jews to immigrate to Israel. This is based in the political Zionist notion of the 'ingathering of the exiles'. Yet not all immigrants are equally valued and, while as Jews they may be important to the demographic project, depending on where people came from, they were differently valued. Jews arriving from North African and Arab states were widely considered to be more 'primitive' and requiring 'civilizing'. Indeed, as in other places, sociology was recruited as a tool of the state for determining connections between poverty and ethnic background (see Dominguez, 1989, for a critique of these studies) in a manner that reified the categories of ethnicity, assumed difference between people and naturalized the *'edot*, 'ethnic groups', along a continuum from uncivilized 'Orientals' to civilized 'Europeans'.

These sociological studies and subsequent governmental policies were aimed at improving the situations of the Mizrahim, Jews whose families came from Asia and Africa. However, they were also often naive and misguided, in ways that parallel some 'race relations policies in the UK and USA, which relied on the assumption of inherent differences among people and anticipated conflict among those 'groups' (Back and Solomos, 2000).

In response to this, on the one hand, there has been a resistance to the devaluation of Mizrahi culture, and, on the other hand, to a growing *desirability* of difference. One instance where this plays itself out culturally is in the question (which I have heard so often in Israel), *'Me-eyeze motsa at?', 'What is your extraction [origin]?'* Meaning at once, what is your ethnicity, and at the same time where did you come from. At the same time, this is often considered to be a rude question. So there is a very specific 'politics of difference' (West, 1990) operating in Israel among people who self-identify as Jewish that has to do not only with whether you are Jewish or not but that is also concerned with geographical movement, skin, accent and many other factors.

This chapter continues the book's overall argument that extracting eggs from women's bodies is a complex technical–material practice, oocyte (egg) extractions make 'the national' and 'the political' palpable. Equally, 'the national' and 'the political' make ova palpable. There isn't just a simple, singular 'egg donation' available for condemnation and critique. Here Israeli notions of ethnicity, in the guise of ideas of 'mixture' and 'difference', are found in egg recipients' selection and rejection of imagined traits and donors. 'Biosocial ethnographic moments' (Nahman, 2007) from the clinics and preferences for traits of donor/baby/self make national borders palpable. They are an enactment of a repro-migration, where the donor is 'over there', the baby 'in here' and the self 'right here'. Distances, both geographic and 'racial', are telescoped. 'Local' and 'Global' become problematic distinctions in all this. Jewish and non-Jewish get 'mixed' by clinic workers for improved viability and vitality. In rejecting Arab eggs, borders and national selves are made and unmade.

5
Borders: IVF @ the Vanguard of Technology + War

Intersecting vulnerabilities

Stacks of fieldnotes from my first weeks in Jerusalem are smudged and cluttered with phone numbers for apartments to view, entry codes for different wards, names of key personnel and phone numbers for different physicians. Pages are folded and half-torn, differently coloured pens had been used. These pages are as chaotic as the days when I began my fieldwork at the height of the Al Aqsa Intifada. While setting up my research site, I searched for a place to live. Meanwhile each day I commuted 20 minutes by taxi between the hospital and the kibbutz (where I was staying for the week) because I was told that the bus may not be safe around these parts because the 'Arabs in the village above have connections with Hamas'. Such warnings were a constant refrain from my interlocutors. These cluttered, chaotic notes materialize 'the field' as a site of intersecting vulnerabilities: my own fear of being caught in a suicide bombing during research,[1] the everyday anxieties of my interlocutors (largely Jewish Israelis with whom I had longstanding and new relationships that were familial, friendly or professional), and a deepening concern for what the Israeli military occupation was doing to Palestinians in the West Bank and Gaza. The complexity of these intersecting vulnerabilities is a much richer story than a narrowly 'domained' account of IVF in Jerusalem could be and it is also a 'truer' story.[2] The 'borders', the individual, the body and eggs were constantly defying artificial separation into apparent 'domains' of analysis (Strathern, 1988), instead there were multiple entanglements (Barad, 2007). What we get then are the synecdochic ricochet effects (Hayden, 1995; Nahman, 2008).

I arrived at the hospital half an hour early for my introductory meeting with Professor Sam Barukh,[3] who would become my main interlocutor.

I sat in the large waiting room for a little while and tried to take it all in. It was very pleasant in this particular building, as compared to hospitals that I had visited in Canada and England (and as compared to hospitals that I visited subsequently in Israel). The floors were a light-coloured marble with blue accents, there were pot lights set high above in the ceiling that are brighter on the edges of the hall than in the centre. There was a glass-encased elevator in the middle of the room. Next to this is a large, round, green rug with big stuffed toys. There were sounds of crying babies and children all around me, which eventually subsided.

As a major Israeli city, Jerusalem is said to reflect the 'mixed' population of Israel as a whole and this mix was present in the hospital as well.[4] Religiously observant Muslims, Christians and Jews and 'secular' people of various backgrounds co-mingled in the waiting room.[5] An older man was speaking English to a young child who responded in Hebrew. There was a small concession stand to one side of this space where I went to buy a fruit tea. The young woman, who was supposed to be selling, was snoozing. I woke her up and she asked me if it took long to wake her up. 'Only in Israel [could this happen],' I thought to myself, echoing words of the self-deprecating opening song to a weekly comedy show, which was then aired in Israel on Friday nights (which much of the country tuned into). An electric toy-train circled the hall about midway up the approximately 7.5-metre high walls. A man was clearing away the toys. It had been very rainy up until then that day but the sun had begun to shine. A Palestinian man entered with three children.[6]

After taking in all of these details I realized that it was time to go and meet Professor Barukh. I made my way to his office, but he was delayed. I waited for an additional half hour, enlisting the secretary to determine his whereabouts and observing others around me. Barukh and I eventually met in the corridor and conducted our meeting in his office. He was an obstetrician gynaecologist, specializing in IVF. Born in Turkey, he now lived in a 'border' area that experiences much violent conflict on a weekly basis. He is 'secular' but observes many Jewish traditions.[7]

Barukh said that he was eager to assist me and began by telling me about the circumstances of ova donation in Israel and at his hospital in particular. I was puzzled and pleased by his eagerness to 'assist'. Having heard from other Israeli academics about the difficulties of gaining access to clinics, surgery rooms and laboratories, I was surprised by the openness of this clinic. Interestingly, other Western anthropologists who have studied in Israel have experienced similar openness (see, for example, Kahn, 2000), conversely some Israeli academics were surprised at the ease with which I gained access. This may be due to the fact that, despite my own 'Israeliness', I was still often viewed as an outsider and my positive reception may have been part of a presentation of Israelis' national self to 'outsiders' as leaders in science and technology.[8] Indeed, this became clear later on. This notion of the assisted ethnography of assisted conception anticipates the rest of this chapter, where nation, nature and geopolitical borders are assisted and reproduced (and see Roberts, 2012, on assistance and IVF).

Over the course of the three months during which I conducted research at this clinic and learned about IVF from Barukh, a rather paternalistic relationship developed with him. He was the facilitator and I had to abide by his rules if I wanted to continue research at his clinic. There were positive aspects to this relationship. When there was a suicide bombing at my local supermarket, Barukh, like a concerned parent, immediately telephoned me to ascertain whether I was safe. Given the conditions of everyday life in Israel, I was grateful for his watchful eye at such times.

Barukh began facilitating my research by explaining the state of egg donation in Israel. At our first meeting, he told me that ova donation at this clinic was at a standstill due to the 'episode that exploded' (Chapter 6). Consequently, Barukh commented, he was ashamed to ask women to donate eggs for fear that they might think that he is involved in a *mashara* or a 'shady business'. He told me that egg donation has become shrouded in secrecy in Israel. He added that the current legislation was unclear and that a committee had ironed out a new law to alleviate the shortage of donor eggs (which came out nine years later in 2010). He explained that he was not pleased with the proposed law because it was too complicated. It involves obtaining permission from the heads of the Obstetrics and Gynaecology units of the hospitals for every single procedure and filling out several forms. In addition to this 'time-consuming bureaucracy', Barukh commented that there is a very low 'success rate' with ova donation. He added that he did not feel that egg donation was worth his time. 'What kind of a physician would engage in this?' he asked. 'One who cares only about money,' he replied. So, Barukh told me, there is not much ova donation to be seen here, but he nonetheless promised to introduce me to women who were waiting for eggs.

Barukh continued to explain that when there was an occasional donation of eggs, the intended ova recipient was required to fill out an adoption form and register her receipt of ova (a similar procedure is undertaken for sperm donation). There is a form to fill out, which is an agreement with the hospital signed by the couple undergoing IVF with egg donation stating that they will care for the child that is born, and that they will make a request with their local court of law to adopt the child that is born thereby. The couple agrees to make this request within one month of the child's birth. Barukh explained that, given that there was no formal body that registered each ova and sperm donation, most couples fill out this form but do not actually obtain a court-sanctioned adoption, because they want to keep their ova donation a secret (see Chapter 4). The then newly proposed law, he feared, would mean a registry of donor offspring that would infringe on the privacy of egg recipients and be controlled by religious authorities. He told me that he feels that doctors should try to prevent this from happening.

After this lengthy discussion Barukh and I talked about the arrangements for my research at his clinic. He was amenable to my request that I be allowed to follow him around the hospital and observe all procedures, from patient consultations to surgery. As we sat and talked, his beeper, mobile phone and office phone were constant interruptions that Barukh told me to get used to. We ironed out the details for my fieldwork at the clinics, including which days I would be observing in surgery and which in the clinic. At this point there was no arrangement for me to observe in the laboratory. Authorization for this lay in the hands of the head of the laboratory, whom I would have to approach myself. Barukh cautioned me that this head would probably not be amenable to my request.

After this, Barukh took me on a tour of the hospital. He showed me around different wards, the medical school and locker rooms (where I would need to get changed into proper hospital greens). He took me to all of the bulletin boards where housing advertisements were posted. Together we collected phone numbers for apartments in neighbouring villages. At the end of our meeting, Barukh handed me a large stack of papers that listed amendments to the IVF laws that pertain to egg donation and religious debates on reproduction. I was to return during the following days to begin 'fieldwork'.

The following day I began research at the clinic, observing as Barukh met with his patients. When I arrived at the hospital, I went directly to the IVF clinic. I asked the receptionist where I could find Professor Barukh. I was told to wait until he returned from surgery. Her terse reply to my tentative inquiry made me feel like a child who had done something wrong and who was being told to 'wait until your father gets home'. Instead of making me feel like an outsider, the secretary's harsh and casual way of speaking to me made me feel as though I was already a part of the clinic's community.

I sat in the waiting room of the clinic for 45 minutes. Couples[9] were also waiting to see him. He entered his office, just four metres away from where I sat and closed the door for a few minutes. The secretary opened his door and said a few words to him. She handed one couple their medical file[10] and informed them that they may enter. While this was happening, Barukh quickly called me in, saying in a commanding tone, '*Shvi. Tavi'i et ha-kisse henna*' ('Sit. Bring the chair over here.'). I rushed to obey, bringing one of his extra office chairs next to his desk to the place that he had indicated. As the couple sat down across from us, Barukh turned to me and said, in English this time, 'You are one of us' (Thompson, 2005).[11]

Over the following hours, and throughout the months of my fieldwork, I would sit in Barukh's office in this manner. I would listen to Barukh speaking with patients about the course of their treatment and try to act as though I was, indeed, 'one of them'. However, sitting in this manner gave rise to some profound anxieties for me. I was uncertain about whether I wanted patients to think of me as one of the medical cohort. I was disturbed by the fact that it was so easy for me to 'pass' as one of the medical cohort (most often as a medical student). Barukh would present me to patients as a student, which would lead people to think that I was a medical student. This troubled me. Yet in the early days of my research I was conscious that I depended on his good will in order to maintain my research position. I made several requests to Barukh to alter his way of presenting me to patients, and, while he did not permit me to use an informed consent form from the patients that I was observing, he did ask patients for permission for me to observe as part of a sociological study of 'the culture of IVF'.[12]

In addition to my ambivalence about being represented as one of the medical team, I often sensed that I was like a 'spy' (and was sometimes told that I was by those I was observing), trying to blend in and not quite succeeding. I would sit in Barukh's office and take notes as patients and physician interacted. Barukh would talk to me about patients' personal histories before and after they entered his office, suggesting that these details may be of 'sociological interest'.

During a meeting with one Jewish couple I stopped writing because the man glanced over at my notebook several times. I was suddenly aware that I had been writing in English. I reminded myself to write in Hebrew when patients (the majority of whom were Jewish) were in the room. I felt alternately invisible and hyper-visible. It was as if the hospital 'greens', once put on, erased my difference, my illegitimacy in the medical cohort. Instances of failure to 'pass' marked my outsider status, such as the occasion when I wore the green top inside out, and a medical student pointed this out to me. In another instance, I rushed purposefully through the surgery ward to the correct room and was stopped by several nurses who yelled at me: 'Hat, hat!' This caught me off guard and I was very confused. After the third nurse yelled at me, I realized that I had forgotten to wear the head covering required in this ward. I was certain that I seemed out of place and would soon have my position of privilege withdrawn.

No one noticed my errors and instead I was granted even greater inclusion. On one occasion I asked Barukh if it would be 'okay' to observe him that day. Having waited patiently for him at his office, I asked my question and he replied impatiently, 'You don't need to call. You know what to do. Go to the surgery room!'

Within the first few hours of my first full day at the clinic I was provided with coupons for lunch in the staff cafeteria, keys to the nurses' locker room and entry codes to the delivery and surgery wards. I was shown where and how to get my hospital 'greens', which I was instructed to wear in the surgery ward (along with a head covering and face mask). I was made to feel as though I was one of them. And yet, throughout the next few weeks I felt very peculiar wandering the halls in search of the locker room, laboratory and cafeteria.

In the early weeks of my research I submitted an ethics proposal (which was approved) to the hospital's Ethics Committee and the Ministry of Health. This application was submitted with the help of Professor Barukh, who presented my study to the Committee. Due to this assistance, during three months of fieldwork in this public clinic I observed countless oocyte retrievals, embryo transfers, Assisted Zona Hatching (AZH), ICSI, egg and sperm washing and doctor–patient consults. I spoke with many patients, interacted often with the nurses and interviewed one social worker. I learned much about IVF. Yet I only interviewed one ova recipient and one donor at this clinic. Their stories are discussed below.

'Making more disposable people': 'the national' brought into the clinic

According to doctors, scientists, government and scholars, the State of Israel is a 'world leader' in IVF and has 'the most IVF clinics per capita in the world'.[13] During my research it was not possible to attend a fertility conference of nurses or doctors, or a panel at an international anthropology conference on Israeli reproduction without hearing about this 'world leader' status. It is like a national battle-cry. At the same time, it is a very emotive statement about the state's borders and history – Israel needs to be a world leader, so the narrative goes, because it is surrounded by enemies (past and present). Israel needs to provide free IVF (a claim countered by many interviewees) in order to support its family and child-centred ideology; a paradox since Israeli IVF is not really all that accessible (Nahman, 2011; Sperling, 2010) and since Israel has one of the highest child poverty rates in the West.

Headed by physicians who are caring, attentive, highly creative scientists and researchers but who are also former high-ranking military personnel, this clinic is an ideal place to see the interplay between inward-facing nationalism of guarding state borders and claims about being at the forefront of technological development.

> Shadiyeh 'Odeh, 24 February 2002:
>
> On 24 February 2002, 28-year-old, pregnant, Palestinian woman Shadiyeh 'Odeh from the district of Nablus went into labour at 4:30 a.m. Her husband drove her toward Nablus to the hospital after depositing her other children with grandparents. On the way they were stopped at a checkpoint, her husband got out to explain to the soldiers they were on their way to the hospital. The soldiers let them pass. A few hundred metres later they stopped to clear a block to the road and other soldiers fired at them, hitting Shadiyeh in the back of the shoulder, the bullet exiting through the front of her chest. The soldiers approached the car and initially did not believe she had been shot. Upon seeing her wound they called the district coordinating officer who facilitated their transport to a hospital urgently. (B'Tselem, 2002c)

I have selected many instances from the research where borders, bodies, nature and the nation are made and imagined in ways that confirm the hegemonic Israeli nationalism at the time of the research. This was a time of deep suspicion of Palestinians, where the 'war' that many Israelis felt that they were fighting was against an Other who could be hiding at any time in their/our midst, ready to explode. This made the everyday running of the clinic a space of suspicion and potential danger.

Yet, most of the doctors, nurses and embryologists who were observed in Israel were deeply committed practitioners. In nearly all of my observations and the accounts of others in this field of Israeli reproduction (most prominently Kahn, 2000), the level of kindness and attention that I observed these practitioners give to their patients was very high and very personal – almost familial. This attention and care did not waiver, no matter the ethnicity, religion or class of patient in the room. Even if behind the scenes they joked as a way of letting off steam, doctor–patient interactions and discussions at team meetings at this and other clinics were extremely caring and respectful.

These undercurrents of tensions around race, borders and nationalism would, nonetheless, rise to the top in moments in the laboratory, the waiting room, the doctor–patient consultation room and the surgical theatre. The almost electric strain of the Al Aqsa Intifada and the IDF's unrelenting 'responses'[14] to it fed to us through television screens would have filled the minds and bodies of the clinicians and scientists. The ride from home to the clinic would have been filled with a deep sense of fear and vulnerability, especially to those Israelis living in the West Bank. Nevertheless, the pressure of being citizens of a state that 'disposes' of its people, as one clinician reported below, has its consequences in moments, hesitations, linguistic turns of phrase and unfunny jokes.

> The *piggu'im* on Saturday night freaked me out. I got scared and used the excuse of Aunt Greta's funeral to leave Jerusalem for a couple of days. (Fieldnotes, 11 March 2002)

These synecdochical connections I make of the ways in which Israeli state policies of invasion and border policing get into the clinic and shape reproduction (whilst reproduction also shapes the nation) are impelled by a strong critique of Zionism and its effects on the lives of Israeli people on a day-to-day basis. The Israeli embryologists, doctors and egg recipient described in this chapter are unwitting victims produced by the material-discursive production of the nation. In the clinic they may not be overtly thinking about the borders but about the bus-ride home. The analysis centres, in part, on how they have internalized that violence.

After each successive patient/couple left, Barukh would provide a commentary on what had transpired. His details involved information about IVF practices, and also seemingly unrelated details about patients' lives.[15] His remarks were often judgmental and/or fatherly and they showed his investment in the success of his 'cases'. When we talked about why the system allows for so much IVF treatment, he explained tongue-in-cheek that this is in order 'to make more disposable people'. He then retracted this statement.

It may be the case that Barukh's 'joke' about 'making more disposable people' is part of the 'unfunny wit' through which people express anxieties about their society's boundaries (Douglas, 1966: 147). Mary Douglas delineates four kinds of 'social pollution' through which a society expresses concerns over its boundaries through humour. Barukh's joke seems to express an anxiety about the fourth kind of 'social pollution': 'danger from internal contradiction, when the basic postulates are denied by other basic postulates, so that at certain points the system seems to be at war with itself' (Douglas, 1966: 146–7). As this suggests, in some cases humour may be used to express anxiety about the contradictions in society's boundaries. Douglas's point may be relevant for interpreting Barukh's 'joke'.

It has been argued that the Israeli state places great emphasis on producing more Jewish people, through pronatalist legislation, media representations and various other practices (Kahn, 2000; Kanaaneh, 2002; Portugese, 1998; Yuval-Davis, 1997). Yet, its expansionist and colonialist politics and military practices (Shafir, 1999) mean that the Israeli state uses its citizens in a manner that 'disposes' of them (through military incursions into Palestinian territory, and by, in this sense, 'provoking' suicide bombings).[16] The term 'demographic race' or demographic war' is an Israeli expression about one of the 'fields' in which the state wages its 'war over home'. One of the major 'threats' to the existence of the state, *as an exclusively Jewish one*, is thought to be the 'over-reproduction' of Palestinians.[17] This has led to many state policies to promote Jewish reproduction so that Jews will win the 'demographic race' and there will be no question as to the religion of the state. Israel's wide support for IVF is part of such policies. (Part of this 'threat' stems from Israel's liberal-democratic parliamentary system of proportional representation.[18])

The state's 'demographic race' against the Palestinians, along with its constant search for new and legitimate sources of immigration,[19] represent two aspects of the attention that is paid to maintaining the 'Jewish character' of the state. These two aspects are contradictory in the sense that the government enacts military practices and policies that endanger the very 'Jewish nation' that it is said to be protecting. Barukh's joke about IVF practices being facilitated by the government in order to make more disposable people may express an anxiety about the contradictions between the state's financial and structural support (in the form of hospitals and programmes) for IVF and its high investment in military activity (recall as well that military service is compulsory in Israel).

The state's concerns regarding the 'quantity' and 'quality' of Jews, the demographic race and the contradictory nature of the state's role in the making and endangering of the lives of Jews in Israel are considered throughout this book. Ova donation intersects in many ways with the concerns about the Israeli nation. Central to ova donation are the issues of the 'quantity and quality' of ova and embryos, as well as the idea of a 'race' (in this case also a 'race against time' due to the age of many egg recipients). The ironic comment, which Barukh made, that Israeli IVF is performed in order to 'make disposable people' is a crucial theme that runs throughout this chapter.

> This morning there were armoured vehicles, military police and barricades surrounding the hospital. Two female soldiers were killed in Be-er Sheva, Qassam missiles were fired into Israel, the IDF fired missiles into Gaza City, wounding 25 people yesterday. Since there was a lot of military 'action' I decided to take a taxi to the hospital. Later that day, Barukh explained to me that there had been a direct threat of attack on the hospital, which explained the presence of armoured military vehicles at the entrance to the hospital complex. (Fieldnotes, February 2002)

Routines of research: 'the national' brought into fieldwork

My weekly routine was to arrive at the hospital either by bus or taxi. This depended on how much violence there had been that week and where it had occurred. Usually if there had recently been a bombing in Jerusalem or if the IDF had gone into a Palestinian city and assassinated someone or bombed buildings, then there was more likelihood of a suicide bombing occurring in Jerusalem. Since such bombings often took place on buses, at these times I would call a taxi to take me to the hospital. On my arrival in the hospital an armed soldier would check my bag for weapons. I would then enter the building and make my way through the maze of hallways, staircases and wards to the area that housed the locker room. There I let myself into the room, found a pair of hospital greens that fit me and got changed. I would then enter the nurses' lunchroom in the labour ward and put everything except my notebook, pen and mobile phone with the bags of the nurses into a locked cabinet.

After getting changed and storing my bag, I would make my way to the surgery wards in the next building. I would be allowed into the ward by an attendant who, on seeing me in greens, would unlock the door and allow me to pass from the patients' waiting room into the surgery ward. Inside I would find another nurses' locker room where there were standard, disposable white head and foot coverings, which I would put on and then make my way to the surgery room by punching in another code on a pad by the two large glass doors. I would enter the surgery room designated for IVF where Professor Barukh was about to begin his procedures.

Victims of suicide bombings were often brought to this hospital for treatment. Co-mingling in the waiting room were concerned Jewish and Palestinian family members of bombing survivors and of partners of IVF patients. It was there more than anywhere else that the contradictions expressed in Barukh's ironic comment became apparent. The routines of my research indicated the fuzziness (or rather impossibility) of Israel's borders with Palestinian territory. These borders were both heavily fortified (with guards, checkpoints and police presence) and impossible to secure, as the 'enemy' was imagined to consist of people living inside and outside those borders. The routines of my research were thus created around both the practices of making new 'Life' and the violence (or threat of violence) of life in Jerusalem.

Learning the social etiquette of speech, gaining insider status

Occasionally, before people in the clinic became accustomed to my presence,[20] someone would question me about my 'right' to be there, and I would respond by letting them know that I was a student and that they should talk to Professor Barukh if they had any questions. I learned throughout my visits and stays in Israel that it is good to be assertive and direct when responding to such questions. What would be interpreted as rude or aggressive in the UK or Canada is just the way things are done in Israel. Apparently my response was never regarded as rude by anyone but myself, and it soon became quite normal for me to respond curtly and be assertive without the niceties to which I was accustomed. One of my favourite aunts in Israel teased me about my former social ineptitude, when she mocked the youthful 'please' I inserted before requests. However, my learned assertiveness worked to my disadvantage when I returned to Britain, where I had to remind myself to use 'please' and 'thank you'. Despite my learning the etiquette of speaking in Israel, there were often misunderstandings between hospital personnel and me that were related to different research disciplines.

Many of my encounters with medical staff at this clinic were met with confused encouragement. Although Barukh had introduced me to all of the physicians in the IVF unit (although not to the nurses, I was left to introduce myself to them), they did not quite know what I was up to, hanging around the clinic, entering offices, carrying patients' files. Still, I was tolerated and occasionally people would ask: 'So, how's the study going?' It was slightly surprising when a month into my fieldwork, while I was sitting in on a unit staff meeting, one physician asked another, 'Who is she again?' and the other replied: 'I don't know for sure, I think she's a social worker.' I approached the second physician after the meeting and explained that I was not a social worker and tried to present my work. I was not certain that he left any clearer about my work than he had previously been.

Although Hebrew is my 'mother tongue', Barukh and I had misunderstandings that were of a disciplinary nature. He often could not understand the reasons for my research. At times hilarious and at other times frustrating, attempting to have medical personnel understand what I was doing and more importantly *why* I was doing it brought up issues of translation, of the 'value' of social research and questions about what kinds of alliances can and cannot be made between those working in biomedicine and those in social science.

Barukh offered to help me translate the abstract for my hospital ethics application into Hebrew. I had thought of doing this on my own but it was too difficult and I did not have the Hebrew facility on my home computer. Barukh said that it was no problem and he would help me. It took us 2.5 hours to translate a single paragraph!

As we went through the form sentence by sentence, I tried to find the correct Hebrew words, and he tried to comprehend what I was saying, we continually reached a road block. First, he did not understand what I want from my research (he did not get 'the point' or the 'outcome'). He laughed at how 'complicated' sociology was and commented on how much he hated this subject at university. He would repeatedly respond to my explanations of the paragraph by asserting, 'I don't understand.' I tried to translate 'kinship' or relationality, and his response was: 'But what does this have to do with what you're studying?' Finally, when I mentioned ethnicity, he said, 'Oh, here we have something,' and he proceeded to write that, 'the study is particularly important because of the different ethnic perspectives on IVF in Israel'. He became mired in his own logic when he realized that ethnicity is also an issue in other countries besides Israel, so he did not understand why Israel was important to study, to which I added that religion is also an important factor in Israel.

We got to a point when I said, 'let's just say I'm doing "ethnographic" research through observation'. He liked this, 'Oh, that's great', he said, 'that'll confuse them and shut them [the hospital's Ethics Committee] up.' His main concern was that the Committee was going to ask him to explain all of the terms of the research, and he was very confused as to what these terms meant. Each time I tried to explain things, he would say, 'Well, that's too in-depth, they won't want to hear all that.'[21]

Finally, we got to the heart of the matter as he declared that no one at the clinic had the faintest idea of what I was doing there, or more precisely why I was doing what I was doing there. They knew that I was studying IVF and them, but they didn't quite 'get' why. When I discovered this I laughed and asked, 'Why the heck would you be letting me in to study you then!?' Barukh's response was, 'Well, it's important to have research done, we're sure you're doing something important. We're just not sure why you're doing it!' He added that the head of department has many requests to conduct research at this top IVF unit and he doesn't have the heart to turn them away. He asked, 'Do you think he has a clue as to what you're doing here? No.'

Yet, I had already explained what I was doing there. I had posted and distributed information sheets and talked endlessly to Barukh about 'science as culture' – I explained to him about the anthropological endeavour being all to do with 'making the familiar unfamiliar'. I had even taken to explaining:

> You know how anthropologists traditionally studied tribes? Well, we realized that our own society is just as fascinating [...] it is equally open to critique and analysis [...], i.e., we want to make medicine exotic, unfamiliar, so that we can learn things about our own cultural assumptions about what is normal, abnormal, etc.

And he still didn't quite 'get it'. I wondered if I should even want him to.

By this point the whole thing was becoming slightly frustrating as I realized that I would probably never make myself understood to my 'primary informant'. Nevertheless, he insisted that my research was important and worthy and that I should not be offended.

He told me a story about a researcher from China who appeared at his office one day. The head of department told him that he could come and conduct postdoctoral research. Barukh hadn't even heard of this man, let alone his project. Barukh explained this case as a matter of international relations: apparently the Israeli army once tried to sell China a Phantom spy airplane but the USA did not allow this because it would endanger their troops, so 'China was offended,' and diplomatic routes were being taken to improve relations between the two countries, so Israel was offering all sorts of fellowships to Chinese academics and entrepreneurs. Barukh explained, 'So, one day, this man, arrives on my doorstep with a project and a strong accent. When you arrived, I thought to myself, "At least this one speaks Hebrew!"'

Rereading these words much later on, I am struck by several things. It is clear from the passage above that although I 'sounded like' an Israeli given my ability to 'pass' in most social settings, I was still viewed as an academic from 'outside', and this made me profoundly conscious that I was being presented with a version of Israeliness that might appeal to the rest of the world.

It led me to question how much medical informants should know about ethnographic research. Was it a 'conversation' that I was having with them, as Rabinow contends in his own research (1999)? Or was this difficulty in translation preferable to complete understanding, enabling me to have the space to conduct my research without being interrogated about it to a great extent?

Barukh was presenting Israel as being a leader in military technology and research. Its relationship to the USA is evident in this presentation. I noted that Barukh was open about the way in which national military concerns enter into scientific and medical practice. It is clear that IVF and clinics like Barukh's become entangled in international politics and diplomatic relations. This makes the comment about 'making disposable people' in IVF seem even more significant.

Exposed: the research hazards of left-wing views

Translation or understanding of one another's views seemed impossible and potentially harmful to my research. By the final days of the period that I spent in Jerusalem, Barukh and I had established a relationship that could be characterized as somewhere between curious engagement and guarded research. For the most part, I kept my emerging political views to myself because of the impact that these might have on the prospect for my continuing research. However, in my final few days I was slightly more relaxed about revealing my political 'identity'.[22]

Professor Barukh kindly offered to drive me and my colleague to the Annual Fertility Physicians' Conference in Tel Aviv. During the drive we chatted about many different topics. At one point Barukh mentioned that his son had just returned from army duty in the occupied Palestinian city of Jenin. He expressed dismay and disappointment at the starvation of the people in Jenin. I was pleased that Barukh was showing emotion about people, whom I had come by now to understand, that he viewed as responsible for the current troubles. I naively expected Barukh to continue and to declare that he was displeased with the brutality of the Israeli military that had (in my mind) caused this starvation. Instead, Barukh's conclusion was that: 'Arafat is a terrible leader.'

According to Barukh, Arafat had caused the current conditions, left his people starving and living in pathetic conditions. Avoiding the question of the politics of the Palestinian leadership, I asked him how this might be the case since Arafat had been confined to his quarters for a few months already. Barukh did not reply. We continued to argue about who was at fault. Barukh maintained that this was all due to Arafat's misuse of funds. He went on to argue that: 'Arabs only listen to force; it's in their genetics.' We argued back and forth and then he quoted the Bible, saying that Ishmael, from whom some Muslims trace descent, was born to the sword. Now he was using genetics and the Bible to prove his point. In arguing against him I exposed myself as a 'leftist'. It was a dangerous move, since I hoped that Barukh would introduce me to more contacts at the conference. The inclusion into the Jewish Israeli collective of any self-identifying Jew (who can prove this identity through matrilineal descent) was still so powerful at that time that I did not need to worry.[23] In fact, Barukh did introduce me to many people at the conference and in subsequent communications he told me that I would always be welcome to conduct further research at his clinic.[24]

Surgical biopolitics

Entrance to the surgery room is gained through a large metallic silver door with a round window. Once you enter this door you are in a vestibule that is approximately one metre wide by three metres long. There is a large metal sink here and a shelf with rubber gloves, pale-green face masks, head coverings and files. At the end of this vestibule is another door of the same size and design. If you walk through this door there is a corridor (same size and shape as the vestibule) that opens up into the main surgery room to your left. In this room, on the right, is a table with a computer that is networked to the IVF clinic's main network, containing a fertility therapies 'expert system' with files for each patient. On the same table are paper files with number codes on the outside and the patient's name, address and telephone number handwritten on the front. On the wall beside this small desk are the printouts for the day's surgical procedures (in Hebrew and English), and an internal hospital telephone. There is a shelving unit with prepacked sterile surgical sets that are unpacked onto a surgical table, and sheets for hospital beds.

If you stand here and look straight ahead there are a heart monitor, an oxygen tank, a stool for the anaesthetist and a metal trolley with syringes containing variously coloured fluids. To the left of this is a black surgical bed with extensions for arms and legs into which the patient may be strapped. Above this bed is a doughnut-shaped light. During surgery this light is on and the rest of the room is dark. The atmosphere is almost of a 'sacred space' when darkened and voices are hushed. This atmosphere is undone when the people in the room begin to converse about politics or tell jokes. To the back of the room is an incubator containing liquids and sterilized test tubes. Next to the bed is an ultrasound monitor. To the left of the room are a temperature control valve and a shelving system with many different kinds of disposable tubes, syringes and needles.

If you turn left again you are at a door to the adjacent room, the laboratory. A small window of about half a metre square connects the surgery room to the laboratory. Access between the rooms is gained through a door that is always open, next to the window. Before each 'embryo transfer' procedure (that is, inserting the embryos into the woman's uterus), just before the physician is about to insert the contents of the tube, the embryologist calls out through the window to the conscious woman: 'Shoshana Blum, Shoshana Blum,' (or any other name) and waits for the woman's reply of '*Ken*,' 'Yes.' '*Be-hatzlaha*,' 'Good luck,' says the embryologist, as a way of legally ensuring the correct identification of the woman and guarding against potential errors.

During a typical 'oocyte retrieval procedure' (one name for extracting eggs from women's bodies), Professor Barukh would sit on a stool in front of the fully anaesthetized patient. The patient's legs were in stirrups, as is common in gynaecological examinations. Behind him was a table with surgical instruments that he prepared prior to the procedure.[25] Usually there were also two nurses in the room (all seem to be Mizrahi and Palestinian women), an anaesthetist (all of them were Eastern European or Russian women in my encounters), a medical resident (male) and me.[26] The oocyte retrieval procedure begins with the sterilization of the room by cleaning staff – who are called in via an intercom system – and with preparation of the surgical tables by nurses. There would be no IVF without this step. Various liquids, syringes and needles are laid out. The physician prepares himself by reviewing the patient's file. Barukh would often give me a short briefing about the patient's IVF history, peppered with personal details about the woman's family life, religious status and her potential relevance to my research. Once the patient – clad in a hospital gown – is wheeled in by two nurses, she is placed on the operating table. The nurses talk to her and ask questions about her medical history. They strap in her arms and prepare her to be injected by the anaesthetist. The anaesthetist asks the woman her weight and then she injects her with a white liquid and as the patient is falling asleep, places an oxygen mask over her mouth and nose. This mask is held tightly for the duration of the procedure, the anaesthetist looks either bored or tired throughout as she holds the patient's head in position and keeps the mask tight on the patient's face, looking constantly at the monitor to check that everything is going well.

It always felt to me as if there was a heightened tension in the room while they were putting the patients to sleep. The women were often nervous and the nurses tried to calm them gently. While the patients fell asleep Barukh disinfected their external genitalia and vaginal canal with a reddish-brown liquid. He used a pair of metal tongs with cotton on the end, which he occasionally dipped, cleaned and replaced. The nurses held a large clear bag into which Barukh disposed of the cotton. The lower half of the woman's body was visible, and a plastic bag was wedged beneath her buttocks. After his careful cleansing, Barukh removed the first disinfectant with a saline solution that was poured directly onto the patient out of a plastic bottle. At this point the lights went down and the doughnut-shaped light was turned on.[27] Barukh inserted a 'transducer' or vaginal ultrasound probe, which has a long (~30 cm) 'aspiration' needle slotted through a metal brace affixed to it. This needle was then used to 'poke' the ovaries and suction up the fluid from the 'follicles'. These follicles appeared as black holes on the ultrasound monitor,[28] and should contain immature ova called 'oocytes'. Once the fluid is drained from the follicles it makes its way via a tube into test tubes that are then passed to the laboratory.

> The past 24 hours have been crazy. Yesterday evening just as I was riding the bus back to Jerusalem from Tel Aviv, a Palestinian man blew himself up in the crowded 'Beit Israel' part of town. Nine people died. The TV news was full of statements like, 'a baby died', 'a woman died'. Nothing was said about the men who died. Am I being asked to feel more for the woman and baby than the man? I feel that such reporting is nationalist propaganda. (Fieldnotes, 3 March 2002)

Embryo transfer: the boundaries of ethnography and the political situation'

While I sat on the bus just as we were re-entering Jerusalem, I knew that something was wrong because the driver increased the volume on the radio and everybody's mobile phones began to ring all around me (a sure sign that there had been a *piggu'a,*[29] and people were checking on loved ones to make sure that they were safe). Indeed my own phone rang. It was my uncle, who told me about the attack before I had a chance to hear the radio announcer clearly.

I considered disembarking before we got right into the centre of town but then decided against walking through unfamiliar parts of the city. Instead I opted to disembark at my usual spot, the rear entrance of the central bus station that is not crowded, and I waited for a taxi. I came home quickly. The whole thing upset me immensely, coming at the end of a week with bombings on both 'sides'.

I went to bed with a headache and could not sleep for a while. Eventually I fell asleep. But at 4 a.m., I awoke to the sound of loud booms – singular, consecutive booms. They continued for quite a long time. The radio and television were reporting nothing. There was no siren to signal that I should go to the bomb shelter (which is a good thing because I did not have the faintest idea where this was located), so I stayed in bed. When morning came, the radio news reported that the IDF had been bombing Bethlehem, which is just beyond the hills that I could see from my study.

That morning I went to the hospital as usual. At first I stayed at the clinic and observed the nurses talking to patients who came in with questions. They took phone calls from women wanting to know test results or about whether there were any eggs available for donation. After observing this for a while I went to the surgery ward. Unusually there was music playing in the laboratory that filtered into the surgery room. The physicians and nurses were talking politics with the laboratory technicians in the other room.

On this day, there were three embryo transfers. I was not wanted in the laboratory and I was encouraged by the physician to remain in my place standing behind him during surgery. To one patient, Barukh said: 'You might feel as though I am being aggressive but I want to remove as much of the *rir* [mucous membrane] as possible so that there are no obstructions to the embryo.' He then went into the laboratory to collect the embryos that the embryologists had prepared and left in a petri dish for him under the microscope. He looked into the microscope and drew two embryos into the syringe. He then returned to the surgery room, where the embryos were inserted into the woman's body.

After observing the surgery I sat and talked to Barukh between patient consultations. We spoke a bit about *ha-matstsav*, 'the situation', a euphemism for the Intifada and the entire political climate at the time, and he admitted that I was there at the worst possible time and he commented that, 'it is very frightening here in Israel'.

At home later that day after observing embryo transfer I turned on the television (for I had become tied to my TV and to constant reports of bombings), hoping in vain that the news programmes will also provide details of what had happened during the previous night's army bombings in Bethlehem, to find that a Palestinian man had opened fire at an Israeli military checkpoint and killed ten people. The news programme included an interview with the father of one of the dead Israeli soldiers. The father stated that his family are proud Zionists and that their son 'died for the whole nation'. This is the language of martyrdom. I became angry that this news report was representing so positively a story about martyrdom when a few days ago it reported a story of a 'proud' mother of a Palestinian 'suicide bomber' so negatively. These different representations seemed to me to involve the Israeli media's participation in a national war of images, in which Zionism was always represented positively and Palestinians negatively. The occupation and its concomitant violence was deeply life-altering and brutalizing for all those affected.

Television and newspapers had started referring increasingly to the possibility of 'transfer'. They employed the English word as shorthand. They meant, of course, 'transferring' Palestinians from the Occupied Territories to somewhere else. One 'option' was 'transferring' Palestinians in the West Bank to Gaza.[30] The language was euphemistic. They were talking about 'ethnic cleansing'. It seemed uncanny that people were using such words as 'transfer' in Israel despite the fact that this was the language that the Nazis used over half a century ago in relation to moving Jews out of Europe.

As I watched television that evening with the news of 'transfer' possibilities, and martyred Zionist men, I could not stop thinking about 'embryo transfer'. The language, so resonant, seemed actually to be referring to things that were worlds apart: 'embryo transfer' for making babies, and the 'transfer' referred to on the news about eliminating the enemy. Yet, they were not as unconnected as I believed that night. The border-making practices of the Israeli state seemed to blur the borders of this ethnography.

Dikkur: the violence of extraction

What is allowed into the Jewish Israeli body, and what is forced out? Extraction of ova is known here as a *dikkur* in Hebrew, or 'aspiration' in English (see Chapter 2 for a discussion of 'aspiration'). The first term *dikkur* is the one most commonly used in the clinic. The word literally means 'a stabbing' or 'a pricking'. When Barukh explains the procedure, he says (in English this time) that he is literally stabbing the ovary. He jokingly tells me that he skewers the ovaries, as if they were a *shish kebab*. In practice he seems very careful, and the procedure looks nothing like a stabbing. Yet this name suggests an internalized violent discourse with regard to getting eggs out of women's bodies. This violent extracting of eggs contrasts with the more 'gentle' popular cultural representations of the practices of 'making babies' that are often concerned with the reproduction of the family, couple, nation or 'global village'. In the case of this clinic, however, this reproduction is seen as a 'stabbing' of a woman's body part in order to extract the cell that will 'make more disposable people'.

Treating the enemy

On one occasion there were five *piggu'im* reported in one day. I spent that day in the clinic in and out of Professor Barukh's office and the nurses' office. I interacted with staff and patients and also observed in the clinic. There were two television screens mounted on the walls of the waiting room, which were always tuned to the news. There in the clinic, when I was trying to concentrate on 'new ways of creating life', I could not avoid the media images of bloody attacks and destroyed buses. Ilanit (the nurse) and I talked for a while about the depressing state of affairs. I asked her whether it changes how she relates to 'Arab' patients. She says 'no way', insisting that she treats them as individuals, giving them the proper level of care. When I observe her with them she is clearly more guarded but very attentive, warm and helpful.

After chatting with Ilanit I observed a *dikkur* from inside the laboratory. Tamar gave me permission to sit and watch what they do on this side. I could hear the patient being anaesthetized, the physician cleaning her and preparing for the procedure. During all of this, while the patient was falling asleep, I overheard the staff in the surgery room talk about where the patient is from, 'Hebron or Bethlehem?' It turned out to be the latter. The nurse said, 'I checked it out, wanted to see if we're neighbours you know. I live in Dekel.'[31] They joked about whether there would be a ceasefire to accommodate the Jewish holiday of Purim and the Muslim *Kurban* (sacrificial celebration), which coincided that year, as the procedure continued and vials were passed into the laboratory where I sat. The sense of vulnerability and borders was bubbling at the surface.

Tamar was at the microscope, determining the number of eggs retrieved. Orly the *mashgiha* was sitting by Nurit and Sarit, the other two embryologists. Tamar was having difficulty finding eggs. So Sarit joined her at the next microscope. While examining the ova through the microscope, Sarit chatted with me about how amazing it is to see the results of her IVF work when she picks up her daughter from kindergarten. At her daughter's school there are the children of couples for whom she has done the laboratory work in the IVF unit. She explains that she looks at these children and thinks, 'I know you since you were under the microscope.'

Quickly pouring liquid from the test tubes into petri dishes, the embryologists checked them under the microscope and found two oocytes. When these were found the embryologists labelled the dish (marking the relative location of the specimen) and stacked up the petri dishes. All the while they were still chatting. Tamar makes a pun with the woman's name: 'What's her name? Shaheen?' There was a pause. 'Shahid,'[32] she mumbled. The other embryologist chuckled. They separated the eggs, washing them of any blood or other fluids. They then manage the *rishshum*, the registration of all of the information on a treatment record sheet that would be copied and placed in the patient's file and the laboratory records. Sarit then told me that these eggs are very small, implying that they may not be 'viable'. 'Eggs must be kept in the incubator for a minimum amount of time,' she added informatively. She then double-checks each of the empty petri dishes and disposes of them. The eggs are moved into the incubator where the space is labelled with the woman's name. She documents the time, number of eggs and the treating physician. The laboratory keeps a file for each patient. If there is an egg donation, this is marked on the sheet. Each egg to be donated is specially marked on the record sheet.

This patient's treatment appeared to be normal and standard procedures were apparently followed. Yet, the talk while the woman was being put to sleep and while sorting her ova was about an enemy. The talk seemed casual, quotidian; there was no air of the unusual. I suspect that the embryologists would not have talked in that manner with me present had they noticed that they were constructing the patient as their enemy. The contrast between Sarit's talk about her daughter's friends at kindergarten (whom she fondly remembers from 'under the microscope') and her talk about the smallness of Shaheen's 'unviable' ova is striking. Tamar's pun on Shaheen's name may be one more instance in which jokes represent anxiety where fear about death have become coupled with discursive practices around borders (the two coalescing into one) and violence erupting in the everyday, make their way into the clinic through such symbolic violence that temporarily repositions a patient as 'the enemy'. The embryologists are of course victims of the upheaval that the Israeli state unleashed on, around and about its borders at this time.

> I got to watch 2 aspirations and a TEFNA [Testicular Fine Needle Aspiration] today at the hospital. I helped Tim, the visiting resident, prepare for surgery – he showed me how to open packages of needles, tubes, etc., sterilely. I chatted with Sandy the Masgikha who talked about not wanting to watch operations – feeling it was an invasion of privacy. Sarit in the lab asked me more about my research. I was independent of Professor Barukh again. I can go wherever I want. So I gave myself a tour of the surgery ward (so I don't get lost as easily next time). Tim explained about the processes of wearing gloves and setting up for surgery. He called me a spy, jokingly, and said that's why he doesn't want to tell me anything.
>
> Soon it will be Purim. Tonight 7 people died in shooting outbreaks near Hevron and east Jerusalem. A Palestinian woman, pregnant, lost her husband and was injured. She was brought here to the hospital. Now she has a baby girl. This seeps into my consciousness without me even wanting it to. How could a cultural account of relationality in Israel not include the daily shootings and killings? (Fieldnotes, 25 February 2002)

The laboratory is divided into three adjoining small rooms. The first room is the largest, where most of the activity occurs. The other two rooms are long and narrow. Each is half the width of the main room. As one enters the laboratory from the surgery room, to the right, there is a table with two high-powered microscopes (see Figure 5.1).

To the left of the window is another worktable with two more microscopes. The work surface is heated to body temperature in order to maintain the 'viability' of gametes while they are being 'washed', sorted and prepared for cryopreservation. Above the work surfaces are photos of infants and 'thank you' cards from parents.

At the top left-hand side of the worktable is a shelf with a cassette player. Around this room are a few large sealed cryopreservation units containing vials of frozen sperm. Additionally, this crowded room has a large refrigerator containing liquids, called the 'medium', in varying quantities for thawing and cryopreserving sperm. The room to the

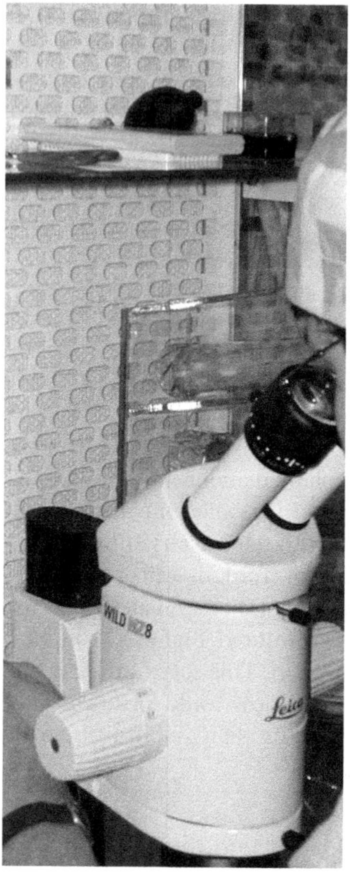

Figure 5.1 Microscope
Source: The author.

left (connected by an adjoining door, which remains open) houses the micromanipulator (see Figure 5.2). This is a large device with two 'arms', which can be adjusted through knobs on either side. It is used to 'drill' a hole into the membranes of ova in order to facilitate the insertion of sperm (AZH) (Figure 5.3), or to insert sperm into ova (ISCI) (Figure 5.4).

Beside this device is a Sony™ video set for enlarged viewing and recording (when necessary) procedures. I was permitted to sit and watch this screen during AZH and ICSI, but the utmost care had to be taken in this room in order not to disturb the very fine adjustments of the micromanipulator. Opposite this device is a desk and shelf unit. The shelf is entirely covered with binders of different colours containing patient IVF records.

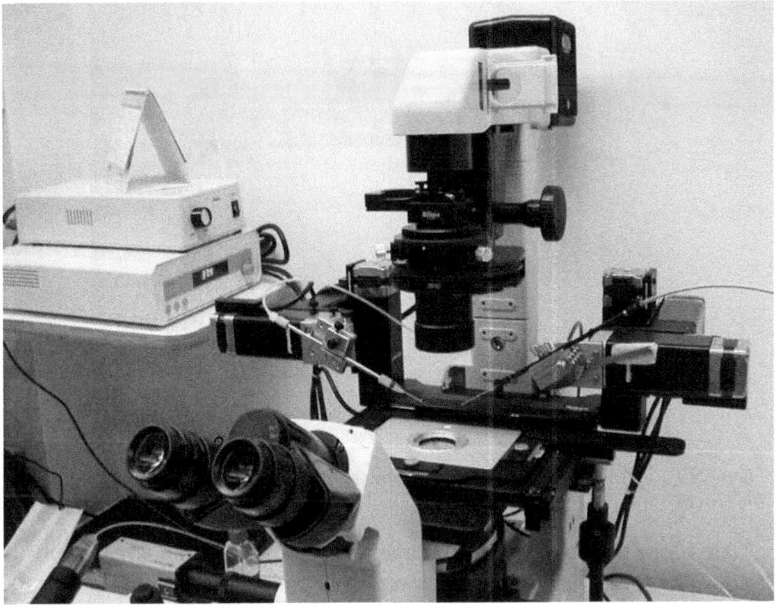

Figure 5.2 Micromanipulator
Source: The author.

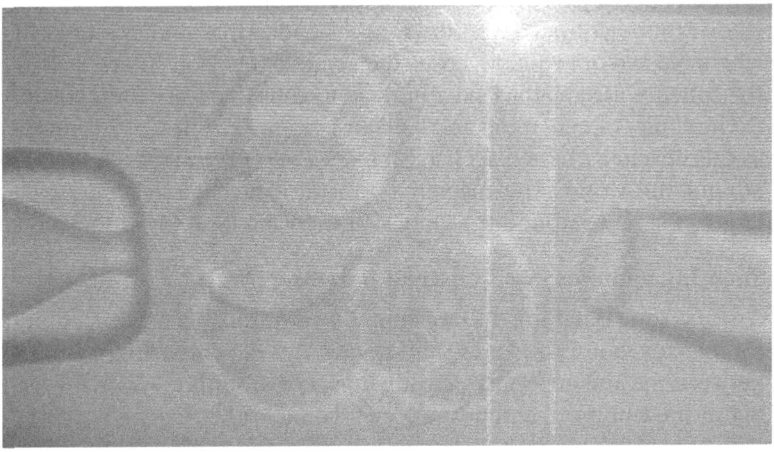

Figure 5.3 Assisted Zona Hatching (AZH)
Source: The author.

Figure 5.4 Intra-Cytoplasmic Sperm Injection (ICSI)
Source: The author.

The room to the right of the large refrigerator houses a large incubator and a cryopreservation unit with temperature gauges. Inside the first are the fertilized embryos at various stages of development. Inside the second incubator are sperm at various degrees of thawing out. The temperature gauges are also very sensitive and connected to an emergency telephone number that dials the embryologists' mobile phones if any drastic changes in temperature occur. On the incubator are small sticky notes bearing surnames of the patients whose gametes are inside. Between the two incubators is another heated workspace. At the end of this room is a large door with a blacked-out window, covered in Winnie the Pooh stickers.[33] This is the entrance used by the laboratory staff to enter and leave the laboratory.

The clinic, surgery room and laboratory are all key sites of IVF practice at this hospital. These spaces are similar to other IVF clinics in hospitals around the world. The machinery, medical practice and science are largely the same around the globe. Yet there are local inflections.

Shooting embryos: Assisted Zona Hatching (AZH)

Assisted Zona Hatching (AZH) exemplifies a cultural scientific technique for making borders, gender and babies. In the following encounter between me and an embryologist who was performing 'AZH', the 'gender of the embryo'[34] and of the border come into question:

Embryologist (E): Michal, sit there where the screen is. What I'm doing now [...]. There are a couple of embryos which will be used soon for *hahzara* [embryo transfer]. Their zona pellucida, outer shell, will undergo a treatment in order to ease their ability *le-hishtaresh*, to take root. What I do is, this machine has a laser beam inside and I shoot, actually on the zona[35] of the embryo and the zona opens a little. And then when he arrives at the uterus, from this area it will be easier for him to get out through the zona pellucida and take root, okay. That's in general. Are you focused there? Can you see what I'm saying?

Michal (M): Yes, yes.

E: So we have here an embryo that has seven cells.

M: Yes.

E: On the left what you see is what holds the embryo in place, and what's arriving from the right is the laser beam. I check that I'm in the correct place, and now I activate her. And I let her to simply [...].

(Successive clicking noises in the background)

E: These are her shots (*yeriyot shela*). And then you will see that the zona of the embryo, in the area that I am touching will no longer be round and beautiful, rather it will be a kind of slit. I don't get into it so that the embryo will be totally open, I just go in a little, so that in this area he will be thinner.

M: And that is so [...].

E: So that the embryo, the cells, when they want to go out, into their *'atifa*, wrapping, [...] it will be easier for them.

M: Okay.

(More 'shots')

E: Now I'll move on to do number three. He's a much more beautiful one, soon you will see. I'll just enlarge so that you can see.

M: What makes a cell beautiful?

E: What makes a cell beautiful? He is round and he doesn't have any breaks. You see on the right-hand side, at around 5 o'clock, there is some small cell breakage. You see the little circles, exactly? Now you see the cell one on the left at the bottom, 7 o'clock, you can see his nucleus very well.

M: Uh huh.

E: So these are beautiful cells, you see their nuclei.

(Clicking sounds)

E: I'm checking, I want a little more. I see that the zona is a bit thinner in this area but I'm still going to try a bit more, you see [...].

(Coughs and more clicks; voices in the background from the next rooms)

M: You're thinning it but not actually breaking it.

E: And I'm not going in. No, no, because then the cells can get out if I break it. Now we'll move to the next embryo.

M: Whoa!

E: You see, this is a very beautiful embryo.

M: This hasn't been frozen yet?

E: No, he hasn't been frozen, he's undergone ICSI. You can see he's got more cell breakage. He's more advanced because he has more cells but he's less beautiful because he has more broken cells.

M: So he has less chance to develop?

E: The same approximately.

M: So what is less beautiful?

E: Less beau [...] embryo that is [...] look.

M: Like, why is he less.

E: A and B both of them give pregnancies, C you see an embryo is full of breakages, that gives less pregnancies.

M: Okay, okay.

(Goes back to 'shooting' clicking sounds for a long time. A phone rings)

E: The first one has more breakages but he's still a very beautiful one. (Embryologist, interview, Israel, 2 July 2002)

In this encounter the embryologist was teaching me about AZH, she was helping me to 'see' what it was that she was doing. The embryo is masculine in Hebrew; along with the description of 'shooting' the embryo, this turns the procedure into something that is not female. At the same time she materialized the border of the egg as a site of puncturing, shooting and taking root. All this is entwined with an ideal of beauty. I was surprised initially when she referred to the embryo as beautiful. I had expected words such as 'good', 'viable', 'strong', but beauty was a repeated refrain here. In the previous chapter, beautiful donors were seen to make beautiful children for parents and the nation. Here in the laboratory, beautiful embryos make good babies.

Injecting eggs: Intra-Cytoplasmic Sperm Injection (ICSI)

I had just interviewed the Palestinian couple, Hamid and Layla, that morning. Little did I know that I was going to see their gametes under the microscope with the head embryologist that day. Here in this interview I watch an Intra-Cytoplasmic Sperm Injection (ICSI) procedure being carried out. The interview bore a distinct aural quality to it, with a sharp pumping sound in the background. I try to convey that sound with the letters 'tsuh, hu, tsu, hu', where the letter 'u' sounds like the first syllable in the word 'early'. The 'ethnographic ear' (Clifford, 1986) here gives an added dimension to the embryologist's description of the process of inserting a sperm into an oocyte:

(Tsuh, hu, tsu, hu)

E: In every micromanipulation technique that you use you usually have two kinds of tools. One is the holding pipette. In most micromanipulation techniques you need to hold either the oocyte or the embryo that are being treated, to hold it in place by a vacuum system that you'll see in a minute, that when you apply it, it just sucks the oocyte to the pipette and eh, that's the way you keep it still, alright?

M: Okay.

E: And the other tool is another pipette, depending on the treatment that you want to do. In the ICSI what you do is to catch a single sperm and to inject it into the oocyte. Okay, you do it in those cases when normal fertilization in IVF doesn't work. Or when the sperm parameters are so low that you don't have enough sperm to use. Right?

M: Okay.

E: So here's a system, you've one to hold the oocyte in place and the other one to inject the sperm. Now what you do now is to prepare the plate in which the micromanipulation procedure will be carried out. This is a plate with a few drops. Within the drops you can put whatever you need. The oocyte the sperm and so on.

M (Recoding explanation of procedure): She's cleaning the surface under the microscope before preparing the plate.

(Can hear another embryologist talking on the phone. Long silence, just pump sound in the background)

M: You're marking it.
E: I'm marking the names. Whenever any biological material is going to be put in, it has to be named. Okay, now I'm going to put now sperm and oocytes and at that moment we have, we have the name of the patient on the plate. The nine, nice drops that you see are medium drops and they are covered with oil to prevent them from being contaminated and from temperature changes. Now I take the medium out, okay. I want two drops. In one drop I will add [gives a name], which is a substance which will help me to reduce the mobility of the sperm in order to be able to catch it more easily. It is a very viscous substance. That's what makes the sperm slow down. Now I look at the name to be sure that's indeed the patient we are working with today. And now I have a sperm. Now let's find the oocytes.

(Opening and closing doors sound. Has a conversation about Mother's Day with a colleague who's leaving. Long silence. Tsuh, hu, tsu, hu)
(We move back to the micromanipulator. Long silence. Tsuh, hu, tsu, hu)

E: We transferred it [the oocyte] to the system. I'm looking at the maturity of the oocyte.

(Long silence, and tsuh, hu, tsu, hu)

M: What's that noise?
E: The cryopreservation machine.

(Long silence, and tsuh, hu, tsu, hu)

M: Okay, so we transferred it to the micromanipulator.
E: We transferred it to the whole system and now let's see what we have, we have nine drops, and this is for someone that has a lot of eggs, so we'll have somewhere to put them.

(Tsuh, hu, tsu, hu, tsu, hu. A sharp, drill-like sound added to this now)

E: Look at the screen and tell me what you see.
M: Okay sperm, lots of sperm.

E: You see a lot of dead sperm or non-motile. But in the edges you see nice moving ones that you can inject, right?

M: Uh huh.

E: This sperm, we don't know exact parameters, but from this picture it would be unable to fertilize in regular IVF, so that's why they need ICSI, although in this case because she's such a low responder and has so little oocytes, it might be a reason by itself [...] usually the rate of fertilization with ICSI is higher. It reaches 80 per cent. So I catch some sperm cells which looked good to me and now I put them in the drop. [...] Okay, what I do now is I try to break the tail of the sperm.

M: Break or grab?

E: Grab and break. I mean, make them nicked. I broke this one and I will go to another one. Okay, make a nick in the tail, it has a lot of biological reasons why, I don't want to go into them but this is a crucial step in the ICSI procedure. Now I put it into the pipette. Tail first, okay and now it's ready in the pipette, you see it?

M: Yeah.

E: Now we go to the oocyte. Now we use the holding pipette and I will hold it. I will have to hold it in a very special position so the polar body, which is a sign of maturity, will be at around between 6 and 7 o'clock and you can see it here now. It is in place. Now you have everything we need. We have the oocyte holding in place by the holding pipette, you see the holding pipette, right?

M: Right.

E: We have the injecting pipette with the sperm in it.

M: Okay.

E: So we're ready for the injection.

M: Wow!

E: You are watching the creation of a new human being.

(Tsuh, hu, tsu, hu, tsu, hu. Phone rings twice)

M: Wow! I've never seen it [the egg] go so flat!

(The embryologist is being called to the phone)

E: I can't right now!

M: What happens now?

E: Uh, I have a very stubborn oocyte.

M: Okay, so it seems to be folding.

E: Hold on, hold on, I need concentration.

(Silence, tsuh hu, tsu, hu)

E: Alright, it is a very difficult ICSI. You see the sperm?
M: Uh, no.
E: It's within the pipette.
M: Ah, there it is, it's still moving. Hmm, that's very [...] [I left off in mid sentence due to her excitement].
E: You see it inside?
M: Wow, yeah, I did.
E: Okay, it wasn't a very optimal ICSI but that's something to do with the quality of the egg which doesn't allow me to go in very easily. Let's try the other one.

(Long silence, tsu huh, tsu huh)
(Embryologist, interview, Israel, 2 July 2002)

As we began the interview, the embryologist had just gotten off the phone to her husband who rang to discuss a problem with their child's summer camp. She then fielded a call for another embryologist from that woman's daughter. This was very much a maternal, familial space, where personal issues were negotiated openly. As the procedure began there was a shift to a much more professional discourse, the embryologist became authoritative and at times impatient with my questions. Her complicated description of the process of ICSI transports us away from a typical maternal environment to one of 'systems' and 'tools'. Whereas I repeatedly named the individual objects, pipettes, petri dishes, micromanipulator, she renamed them 'the entire system'. ICSI comes across here as a collective agglomeration of objects and processes that make one entire whole 'system'. It may be that this is how all embryologists refer to the procedure. Yet, this also bears an ideology of a collective. At the same time, the Palestinian woman's oocyte is figured as stubborn and not easily penetrated. The whole interview can be read as synecdochical to the complicated national struggles and fears that mix the personal, the private, the microscopic, the geographic, with the state. The sound of the cryopreservation machine repeatedly 'tsu, huh'ing' in the background only adds to this narrative.

Borders and bodies

It was in the early days of my research in Jerusalem that I encountered ova donation for the first time. I awoke late one cool, wintry Jerusalem morning, since I did not hear my alarm. I was not sure whether I would become accustomed to waking up at 5:30 a.m. everyday to make it in time to watch those early morning egg retrievals. In any case, this morning I did not make it and had to call a taxi to take me to the hospital.

On my way to Jerusalem the other day I thought to myself, 'how ugly this all looks'. On previous visits I had experienced coming here as something mystical, a pilgrimage. This time I noticed only scrubby, dry hills, broken-down old houses and decaying military tanks that had been left along the Number 1 road to Jerusalem, remnants of the battle that 'liberated' Jerusalem from the Jordanians. The road to Jerusalem looked ugly and depressing to me. I was constantly thinking about the fact that the army just re-entered Jenin, and that they had killed many Palestinians (the numbers at the time were unknown to me, but there ensued a series of investigations by Human Rights Watch, Amnesty International and the UN; the claims and counterclaims as to the numbers killed are highly ideological and it is difficult to assess the truth in such cases); that a Palestinian man walked down Yafo Street and blew himself and 20 other people up. There's nothing mystical about it. I resented having to be here.

The taxi driver, Haim (meaning 'Life'), was friendly and the hills of Jerusalem were covered with cypress trees and a mist, which restored them temporarily to the glory of my childhood memories. Haim asked me if I worked at the hospital. I said 'yes', in order to keep things simple; so that I could continue to enjoy this view quietly. 'Which department?' he inquired. 'Women's,' I replied. He was not to be deterred. 'My wife has been going through fertility treatments for four years,' he stated, and I was stunned out of my reverie. Incredulous, I thought that I had imagined what he had said. However, I was not imagining anything and it seemed that the more I probed, the more Haim wanted to tell me.

He explained that he and his wife had been undergoing fertility treatments for four years. They had gone to several clinics and through many cycles of treatment. None had worked. 'It is very difficult,' he said. He tells me that they *even* went to Romania to get ova through a physician in the Tel Aviv area, but to no avail. They are still hopeful, Haim tells me and he explained that he must persist because, as he says, 'we have no choice'. This idea of having 'no choice' but to undergo IVF and ova donation has been commented on by others researching assisted reproductive technologies (see especially Franklin, 1997). There are some ways in which the context of Israeli IVF makes it a particularly interesting comment as it is interwoven with accounts with reference to the war over the homeland, and links to the reference to 'making more disposable people'.

I was surprised that Haim was so forthright. Since I had arrived in Jerusalem a week earlier, I had been warned that no one would speak to me about ova donation because there was such secrecy around it. Nevertheless, here was Haim, divulging all of the details of his experience with it. We talked and talked until we arrived at the hospital. I exchanged telephone numbers with Haim and gave him my project information sheet to pass on to his wife, in case she wanted to be interviewed. I rushed, breathlessly, into the surgery ward and found that Professor Barukh was not there today, there was no surgery, no clinic. He had made a mistake when he gave me the details yesterday. I wondered if this had been some sort of mystical Jerusalem experience through which I was 'meant to' meet Haim. I then chalk it up to coincidence.[36]

> Maysoon Salah Na'if al-Hayek, 22 years old was travelling with her husband and father-in-law to a hospital in Nablus from a village in the surrounding area. She was nine months pregnant and in labour. They were stopped at a checkpoint for more than an hour and involved in heavy argument about going to the hospital. She was examined by soldiers seeking proof of her pregnancy. Finally the soldiers let them go but a few hundred metres later they were fired at heavily by the soldiers. Her husband and his father were severely injured by IDF bullets and she was hit by shrapnel. The soldiers came to examine them and undressed her to see if she was hurt. She was left naked while they called the district commander. They transported her to a hospital and she gave birth to a healthy baby girl. At the time that she provided her account to B'tselem, she still did not know what had happened to her father-in-law and husband. (B'Tselem, 2002d)

Conclusion

It used to be the case in Israel that gestation (carrying a baby) and not genetics was what decided Jewish kinship, and hence Israeliness, from a legal and religious perspective (Kahn, 2000; Feldman and Wolowelsky, 1997). We have seen already in previous chapters that Israelis themselves perceived this slightly differently and were often ambivalent about the status of the donor, her eggs and what they conferred on their future, potential children. Since 2010, it is illegal to donate and receive eggs between women of different religions. Individual women's bodies are increasingly sites of negotiation of the border, national bodies, citizenship and the state. In the past year, Israel has threatened to deport about 400 children of migrant worker women (Ravid and Weiler-Polak, 2010). As I write this, the Israeli state has begun the process to deport one Israeli-born child, the four-year-old daughter of a Filipina care worker (Paraszczuk, 2011). The creation of borders through blockades and transfer is in this sense a reality now. So the time and geopolitical space of the research can be seen to materialize both egg donation and the border.

Whilst I was conducting this research, during the height of the Second Intifada, the 'apartheid wall' between 'Israel' and the 'West Bank' was in the early phases of being built. These words are in inverted commas because these are not two distinct, easily traceable places. This wall is actually a series of walls and fences, which run along the 1967 border. This border is constantly being drawn and redrawn. Despite the amount of funding that was put into the building of this wall, and all of the military personnel required to staff the checkpoints along it,[37] this wall did not act as a real deterrent to entrance to Israel. From speaking to young Palestinians and international activists, and from my own experience, it is possible to traverse this weblike 'border' if one so desires.[38]

Disposability and transferability of ova through IVF have repercussions for gender, which like borders and eggs, must be generated and regenerated. It is a site of rupture and reaffirmation.

When I returned home, I wondered how it could be possible for me to write a book about a 'life-making' technique, when my research was set at time of so much death in Israel and Palestine. Barukh himself tied life and death together by saying that his government was providing IVF in order to 'make more disposable people'. At the time that he said it, people were dying daily in Israel on buses, in town centres and in synagogues. His ironic comment may indicate that he sees IVF as a way of fostering the creation of 'life', compensating for the numbers of people dying. Yet this perspective seems, once again, to invoke Israeli victimhood while occluding any consideration for the numbers of dying Palestinians. The idea of 'making life' to compensate for the deaths involves Israeli national imaginaries that materialize Israel's situation as that of 'victim'. 'Making more disposable people' could be viewed in another way.

Rather than expressing a solely psychological response of one person, this joke can be seen as a 'material-semiotic generative node' (Haraway, 1997) or a way of doing theory through the lay formulations of people (Strathern, 1991; Verran, 1998). This is fundamentally about the question of which life has greater value to the state. Notions of waste figure centrally to the classical anthropological canon of so-called 'primitive' societies. The Israeli framing of Palestinians and some Jews as more primitive and therefore expendable fits in with this literature in the sense that certain kinds of pollution, sacrifice and waste are viewed in classical anthropology as ways of performing social relations and consolidating them (Mauss, 1990 [1950]; Douglas, 1966; Titmuss, 1970; Hubert and Mauss, 1964 [1899]).

Arguably then, the Israeli state is attempting to consolidate its status as a Euro-American power and this is being done not only through the management of the population, but also through the disposal of 'bare life'. In other words, this notion of making more disposable people can be seen as a discursive moment that indexes biopolitics and the sovereign power to dispose of life. Focusing on Western history, the sovereign has had the power to take a life (execution without sacrifice, as is outlined by Agamben in *Homo Sacer*, 1995). Foucault (1978) argues that in the new era of biopolitics, there is greater emphasis on the management of life (the population and of bodily discipline). Agamben (1995) extends and critiques Foucault by insisting on the centrality of the disposability of life and the creation of death in contemporary biopolitics. He does so through a notion of a juridical 'state of exception' (borrowed from the work of Carl Schmitt), in which life is made legally expendable by making it extra-legal (see Butler, 2004). Agamben (1995: 6) argues that 'psychological' interpretations (such as Douglas's, 1966, perhaps) are insufficient. Something more is going on here with Barukh's assertion that IVF is about 'making more disposable people'. IVF is always about producing an excess of embryos. The processes of ovarian hyper-stimulation, ova fertilization, freezing of embryos and their implantation involves disposal at every stage. So in this case it is about disposal of 'potential' people, what Thompson has called 'promissory capital' (Thompson, 2005). If a woman's ovary produces 20 ova, not all of these will be seen by the embryologist as 'suitable' for fertilization; they might be seen as 'too small', or their membrane 'too permeable' or 'soft'. Hence disposability is an integral part of IVF in general. In Israel this disposability is directly linked to the politics of maintaining the state and imaginaries of the nation. Seeing these connections means attending to the ways in which 'the realm of bare life gradually begins to coincide with the political realm (Agamben, 1995: 9).

The intersection in this research, of watching the processes of 'making life' and finding strategies of avoiding death, was so intense that I cannot think about IVF in Jerusalem without thinking about the deaths of hundreds of civilians in Israel and Palestine. In this sense, with its support for IVF programmes, and continued military actions against Palestinian cities, the Israeli state does appear to be helping to make 'more disposable people'.

The discourse of world leadership is one that chimes well with the notion of the Israeli state as 'the only democracy' in the Middle

East, and a centre of true enlightenment in the 'East' (read: primitive, despotic, closed). There is no doubt that Israel 'leads' globally in many areas of science and technology. The repetition of this idea, however, also begs the question of the ways in which the state does not 'lead' or where perhaps its leadership is in areas that are not so desirable. For example, Israel is among the countries with the highest multiple births rates, something that is attributed to a high proportion of fertility treatments (Even, 2011), which means that there is a greater danger of maternal and infant morbidity and mortality.

Another area in which this 'leadership' occurs is in science and technology, an area in Israel that is intimately bound up with the military industry and the military. High-ranking scientists and researchers are very often former high-ranking military and intelligence personnel. Many Israeli technologies also started life in military intelligence and later translated into biomedical and other technologies (for example, a capsule containing a small camera that can be ingested by a patient for diagnosing internal problems without the discomfort of endoscopy). The transition of technologies and personnel from military to medical, computing and other arenas has been so successful that Israeli science and technology is regularly called upon, globally, to participate in research. In one distinctive arena, that of security and surveillance technology, Israel has become a homeland security global capital and is the sixth largest exporter of military technologies (Gordon, 2011). In part, such success is due to the 'collaborative public space' enabled by the military, where Israeli scientists collaborate and share formal and informal knowledge in an environment that is highly stimulating (financially and otherwise), promoting technological development (Breznitz, 2005).

The financial provision of such development also has a transnational dimension. Until recently the USA was the largest financial supporter of R&D in science and technology in Israel. Most of this money goes into military science and technology 'focusing on computer and electronic technologies, electro-optics, aeronautics, mechanical design and metal works as well as chemical and software engineering' (Gordon, 2011: 155). Shifts in 'civilian' research and development have meant that Israel is, more and more, participating in the European Research Area (ERA) (Zimmerman et al., 2010).

These economic ties with Europe, being a leader in the world of science and technology, and the inward-looking requirements of Zionism (the cultural imaginings of Israel as a Euro-American space with preference for racially European and 'improved' babies, the shutting down and hyper-management of borders) place most Israeli subjects in an extremely vulnerable position.

6
ExplosionCrisis: 'The Episode that Exploded', Women's Bodies and Nationalism

This is a story of a crisis, a crisis of a woman, a group of women, of eggs, and of a nation-state struggling to maintain its state-ness through a dialectic of victimhood and violence. This can be seen as a singular crisis – refracted and repeated differently, multiple-ly. To see them as one singular crisis is to make a deliberate analytical slippage, for the sake of critique. There will no doubt have been many non-parallels between these two stories – of course there were. I am picking the bits that are similar, to think through. I am writing here in the realm of ethnographic imagination. Where does complexity and contradiction in the data mask or make inaccessible the ground for moral critique and interference? I began in Chapter 1 to discuss a feminist methodology of interference. Here I continue this endeavour in seeing the work done by notions of victimhood and crisis in the mutually constituting spheres of the egg and the nation.

The unknown donor

According to Israeli and international news media accounts, in May 1999, a 31-year-old woman began fertility treatments at an Israeli hospital. Despite the fact that the accounts almost always begin with the doctors who stole eggs, my retelling locates this woman at the centre of several interconnected narratives and prioritizes her story above those of the doctors. Unfortunately not much more than I have written here is publicly known about this woman whose identity is obscured in the news reports of her case.

Being Jewish, relatively young and involved in IVF treatments, she was seen as the 'perfect donor'. Her physician asked her if she would donate some eggs.[1] The woman initially refused on 'religious grounds'.[2] She later gave in to the doctor's request and agreed to donate 5 or 6 eggs. As journalist Judy Siegel-Itzkovitch suggests, the woman asserted that on 14 June 1999, the physician extracted approximately 60 eggs, 30 of which were 'donated' (Siegel-Itzkovitch, 2000). A few days later this woman was diagnosed with what is known as ovarian hyper-stimulation syndrome (OHSS). Her ovaries were over-stimulated, which caused abdominal pains and difficult breathing. After being treated for this usually uncomfortable and occasionally life-threatening condition, the woman lodged a complaint against the physician with the Ministry of Health. The doctor, who was employed by a state-funded public hospital, also had a private clinic where he had treated her. Her eggs had been sold to women who were paying for them themselves since private egg donation was not covered by the various medical insurance organizations. The implication was that this doctor had shifted his patient to a private hospital in order to be able to sell her oocytes rather than transfer them at no cost or profit to another woman within the remit of the public health system.[3]

Once these activities were made known to the police an investigation began. Two physicians were said to be involved in a network of 'egg theft' spanning several public and private hospitals. They were investigated under criminal charges of 'ova theft' and negligence (Reznick, 2000a, 2000b; Kraw, 2000a, 2000b; Siegel-Itzkovitch, 2000). They were put under house arrest and accompanied by police officers when they needed to visit patients urgently (ibid.). A governmental committee was set up to investigate the two physicians in question. However, no other physicians would come forward to provide testimony against the two alleged offenders.[4] At least one of the doctors was eventually convicted and removed from practice for a period of time. He is back practising fertility medicine as a leading specialist.

Due to the fact that some physicians had stolen ova, others were now afraid of being seen to be engaging in *mashara*, 'shady business',[5] and were therefore not asking women to donate eggs. In addition the newspaper reports and physician accounts began to frame IVF patients as 'selfish' for not voluntarily donating ova. They were also now being represented as afraid to donate for fear of ovarian hyper-stimulation. This was seen as leading to an 'egg shortage'. The media reported, and later my research participants repeatedly told me, of the fact that in order to remedy the shortage and prevent further theft, new legislation was being drafted so as to allow a more liberal donation of eggs. At the same time, several Israeli physicians began to set up transnational egg donation services either in conjunction with other clinics or by setting up their own clinics in Cyprus, Romania and other parts of Eastern Europe.

Reiterative crisis in Israel and the persistence of 'emergency'

Crisis in Israel is reiterative, maintaining a state of 'emergency'. The State of Israel is so often narrated and performed as being in crisis, that 'crisis' comes to seem banal. Indeed 'banal nationalism' (Billig, 1995) in some states depends on crisis. Whereas for Billig this was the case for 'established nations', here we can see this banality in a relatively new 'nation'/'state'. In an interview with a leading gynaecologist who was also a religiously observant Jew this became salient.

> I was sitting with Dr X in his home having a chat about egg donation ethics and policies today. In the middle of our interview, his mobile phone rang, and he answered it. It was someone from the military calling him up to army reserve duty. He and the person on the phone engaged in a discussion about military matters. I was fascinated and slightly shocked that I was privy to what seems like a kind of military strategy discussion during an interview about egg donation! He talked about the battlefield, 'quieter areas and more active areas' of fighting. He was receiving instructions of where to go. He mentioned different kinds of army personnel, requesting some notification for when to be ready. He asked the person on the phone not to tell his wife that he himself agreed to go, rather she should be told it's a command from above. When he got off the phone he told me he's in waiting to be called up in case Hezbollah wants to start a war with Israel. He finished off our interview a bit later claiming that I received some non-conventional answers from him. I began to wonder whether a full-scale war was about to erupt. (Fieldnotes, 24 June 2002)

In the early days of fieldwork (2002), an Israeli anthropologist heard me talk of the 'crisis' that had emerged in Israel with the theft and subsequent shortage of eggs. 'What crisis? We have forgotten it already!' she said. There were so many crises that this one had already been pushed to the recesses of cultural memory. Yet, precisely this negation of it, 'we have forgotten', with the invocation of a collective 'we' may signal the uncanniness of crisis in the Israeli national imagination (Freud, 1985 [1919]). Indeed, Israeli newspaper reports published years after the events contradict the 'we have forgotten it' argument. An article in the centrist-left national newspaper *Haaretz* from July 2008 cites the 'problems' in Israeli egg donation as stemming from the 'eggs scandal' that 'exploded'

(Linder-Ganz and Darom, 2008). As I indicate below, the repeated refrain of the 'episode that exploded' (referring to the media explosion of the idea of the theft of eggs) and its negation, hint at the 'interminable uncanny' (Gampel, 1996). That is, they hint at the persistence of the fear of violence and state rhetoric that reinforce that persistence.

> I didn't go to Yosef's *Shabbat hatan* [the Ashkenazi term for the practice of the groom being called up in synagogue to recite a blessing from the Torah during the week following his wedding today] due to family pressure to stay away from large gatherings of people. Doda Greta died on Thursday, I go to the *shiva* tomorrow. It's becoming too much to handle – waking in the middle of the night to sounds of bombing and helicopters [the IDF bombing Bethlehem and Palestinian villages near Jerusalem], being scared to go anywhere, everyone around me thinking the total opposite from me [that is, me being more in support of Palestinians than the people I meet].
>
> Two *piggu'im* tonight. The first in Netanya – shooting into a hotel and the second here in Jerusalem on Rechov Azza right near the Prime Minister's residence, 10 died and 50 were injured. (Fieldnotes, 9 March 2002)

I intended to start research in Israel in September 2001. However, just a few weeks before my scheduled departure date (13 September 2001), a series of bombings occurred in central Israel, killing many people. These bombings occurred in the early days of the Al Aqsa Intifada against Israeli military occupation of Palestinian cities, towns and villages. The Israeli government responded by increasing its already strong and violent military presence in those cities, towns and villages. The bombings were also framed as 'a crisis', which put the nation in a *milhama 'al ha-bayit*, 'war over the home' [meaning a war over territory].[6]

> Arrived home around 4 p.m. Talked to Gabby and decided not to go to her parents' house with her. Around 4:30 p.m. there was another suicide bombing on King George Street – it must have happened as I got online to check my email. An eyewitness described the scene – she observed 2 'black' Arabs in a car with blue plates. Apparently this is the 4th suicide bombing attack this week – this seems strange to me – I must have been really tuned out this week because it didn't feel like there were so many – have I become desensitized? (Fieldnotes, 21 March 2002)

When I finally arrived in Israel in January 2002, after many delays due to the prevailing violence and fear during the Second Intifada, people told me repeatedly that 'this is the worst it has ever been'. Jonathan Boyarin observes the same narration of crisis in, *Palestine and Jewish History: Criticism at the Borders of Ethnography* (1996). His research was conducted during the First Palestinian Intifada, in the early 1990s. In reading Boyarin and talking to Israelis, I had the sense that 'crisis' is narrated as an ongoing way of life in Israel.[7] Certainly there was a constant stream of newspaper and television reporting of 'crises' (from economic, to political and military) during the time that I was there.

> Another *piggu'a* today – this one in a Haifa restaurant. 15 people died and around 30 are wounded. And another *piggu'a* in Efrat (in the territories). Arafat has been isolated in Ramallah. The army has taken control of the town – suddenly 40 international and Israeli anti-occupation activists get into his compound – kissing him, showing solidarity [...] Last night there was a *piggu'a* on Allenby Street in Tel Aviv and two days ago there was one at my local supermarket. We are waiting to see what happens in Arafat's compound. If he's killed it will be terrible news for Israel. The violence will escalate unbelievably. That must be what the government is planning. Otherwise why would they call an extra 30,000 *miluimnikim* [reservists] out? I know several people called. The mayor of Haifa speaks and calls for continued co-existence – a political solution and not a military one. My father telephones, as do several family members. Everyone urges me to leave Israel. (Fieldnotes, 31 March 2002)

The history of Jewish oppression and specifically the Shoah is employed by the Israeli state as a 'discursive instrument of political legitimacy' (Ochs, 2006: 359), granting the moral correctness of the occupation over Palestinians and European Jewish supremacy over non-dominant subjects (Shohat, 1997).[8] Crisis was a part of the Jewish experience in Israel during this time. An average of five suicide bombings per week occurred in Israeli cities during the most intense months. These were in reaction to, and often swiftly followed by, Israeli military violence in Palestinian civilian areas. Media coverage was constant and the radio reports diligent. These reports repeatedly framed the State of Israel as oppressed. Jewish people who argued that there needed to be attention paid to the ways in which Palestinians were suffering under occupation were seen as mutinous.[9] Rhetoric about this 'state of emergency' by the Israeli government and media seemed to perpetuate the notion of Israel as 'oppressed'.

This was most obvious in my daily interactions during fieldwork. In the course of doing research I was also living in Israel, among friends, family and colleagues. I had repeated conversations with them about the contemporary state of violence in Israel and Palestine. In most conversations people tried (unsuccessfully) to convince me that Israel was the sole victim and that Palestinians were the sole oppressors, against whom security measures had to be taken. There seemed to be no option for thinking of the suffering among Jews and Palestinians alike.

Despite this, one Israeli lawyer whom I spoke to commented to me that the fundamental legal problem with the state was its Jewishness. He contended that in order to exist as a state it had to remain a Jewish state. However, it seemed clear to me that in order to remain a Jewish state it had to ensure the reproduction of Jews and the suppression of Palestinian rights and reproduction (this has been argued at length by scholars such as Kanaaneh, 2002; Portugese, 1998; and Yuval-Davis, 1997). Palestinians had to be discursively rendered an absolute enemy who wanted nothing more than to kill Jews and take over their land. The lawyer quickly followed this with his view that Jewish Israelis saw themselves as being in a crisis, 'a war over home' (and that I should be loyal and join in this fight). The overriding preoccupation of Israelis I talked to was with the continuation of the Jewish people and the State of Israel *as a Jewish state*. Significantly, it was in 2002 that an intense governmental preoccupation with the Jewish demography of the state was initiated with the reinstatement of the Center for Demography, which was only devoted to the maintenance and study of Jewish demography in Israel. This preoccupation of Israelis and the state with the crises of demography and the homeland was couched in the language of 'victimhood', which in itself is a vehicle of nationalism (Berlant, 1997). The suffering of Palestinians expelled from their lands in 1948 was rarely mentioned in daily discussions, newspaper or television representations of the present conflict.

> I've returned to Toronto for a few weeks to help me detach a bit from the *'matsav'* [situation]. Since I've been back I attended an educational evening led by my friend's activist group – a coalition of Jews, Israelis and Palestinians working to dispel myths about the Barak peace offer. They divide us up into smaller group discussions. A Jewish woman invokes anti-Semitism and the Holocaust as two reasons why 'Jews need an army'. She also said this evening was too 'pro-Palestinian'. Another woman in the group thinks it's too 'pro-Israel'. (Fieldnotes, 30 April 2002)

Alhough emerging from the arena of reproduction and the bioeconomy, notions of victimhood and crisis also made their way into my research through the narration of the 'episode that exploded'. This is a parallel that only I am making; occasionally participants would make the link between Jewish victimhood and the need to reproduce, but the connection between a crisis of egg theft and a crisis of the nation is one that I am making with the aim of drawing an unlikely juxtaposition. The aim is to think about and analyze reproduction and the bioeconomy war and imperialism, through one another.

Michal: Okay, I have lots of questions. You can begin wherever you'd like. I want to know how you got into all this, where did you start [...].

Dr Shmuel: It's simple. We had a crisis in egg donation in Israel. And for us, in our cultural sphere there were quite a few women who needed egg donation and [...]. We didn't have a source [of eggs]. You are familiar with the Israeli law, which only permits us to take ova from women in IVF [...]. (Dr Shmuel, interview, Israel, 11 July 2002)

*

Miriam: And then the episode exploded with the doctor. I had opened a file with him. [...] He was involved in the story and it was already when the episode was at its peak and [...] but they didn't know what its direction would be [outcomes] and he told me it would be alright and I don't need to worry and I will succeed and all. That's it, that was in fact the only meeting I had with him and afterwards the episode exploded. That's it, and everything stopped. And I was [...] because all in all time is working to my detriment [...]. (Miriam, Jewish Israeli egg recipient, interview, Israel, 3 September 2002)

*

> What that doctor did, and the judge, and it exploded. (Yossi and Sarah, Jewish Israeli egg recipient couple, interview, Israel, 2 July 2002)

*

Nearly every single time that I began an interview I would be told about the series of egg donation crises, 'the episode that exploded', until this reference to explosion seemed banal. Recall Shoshana, the 49-year-old egg recipient discussed in Chapter 4, who narrated her being at the private egg donation clinic as a matter of 'no choice'. She had been treated in Herzliya, where the scandal of the theft of eggs had erupted and so she came to this other clinic, because she didn't have any choice, 'The tap was closed.' She explained that there was at the time a black market in eggs. Furthermore, she argued that no one in Israel cares that the doctors who were accused (of the theft) were still working. For Shoshana, she was 'making a war in order to have a child'.

*

Dorit: So [...] and about the egg donation [...] and then I already knew that I am in a very bad problem, because then I knew that already, a year ago the thing was the egg selling and everything, I knew about egg donation. I hoped that I would not have to deal with it. But unfortunately [...] one of the reasons [...] also [...].

Michal: You knew about the whole crisis [...].

Dorit: Yes, I knew about the whole crisis. There is a big crisis. There are no egg donation in Israel today, because of that situation, because women were reluctant to donate eggs. Very [...] one here, one there [...] there is a very long waiting list [...] in Assuta or whatever. No egg donations in Israel right now. Women are [...] again, reluctant to do it, because I don't know. I really don't know. One of the things that I knew, at least the one that I heard, that everything is from gossiping and newspaper, that women [...] that he gave women hormones to have hundreds of eggs! Which is not human, it's not human. I think, I am not sure, that a fertile person in her twenties or thirties, in a good round, she will have 30. Not hundred! Right? So [...] 30 is a normal for a good woman. Good? Whatever [laughs]. Hundred is non-human. (Dorit, Ashkenazi Jewish Israeli, interview, Israel, 6 March 2002)

Even before I went to Israel and in the early days after my arrival, both egg donation and the national struggle for existence were framed as 'explosive'. The theft of ova was referred to by clinicians, patients and in the newspapers as 'the episode that exploded':

> The chapter of the selling of ova is one of the severest/gravest in the history of the health system in Israel; so said the heads of the hospitals in the Ministry of Health administration yesterday. (Reznick, 2000a)
>
> the episode exploded. (Reznick and Shadmi, 2001b)

This language seemed odd to me at the time as the everyday bombings seemed to be about a much more serious kind of explosion. In the year prior to these explosions, the newspapers rang with reports of the theft of eggs and led to a crisis of a shortage of eggs in Israel.[10] These accounts of crisis and subsequent shortage were followed by reports of a renegotiation of Israeli legislation on egg donation, and also by accounts of initiatives to import ova to Israel (Fishman, 1998, 1999a, 1999b, 1999c, 2000; Rozenblum, 2000c; Gross, 2001; Shmerling, 2000; Milner, 2001; Shteiber, 2002; Siegel, 2001). This series of events was recounted to me by clinicians and lay people as, 'the episode that exploded'.

Many times I overlooked these 'crises', thinking that they were irrelevant. Certainly the bulk of my interviews did not focus on the crisis. They merely began with it. It was the trope with which interviewees began to narrate their egg donation story. The remainder of the interviews were generally quite typical stories of infertility, familial pressure and support, financial hardship and the limits and extent of governmental support (as has been shown in previous chapters). However, given the narratives that prevailed about war and crisis during my research, and my own experience of being in crisis over whether to continue my research during what seemed like a war and what was a genuinely dangerous time, the accounts of the ova donation crisis are significant. They share some aspects of their grammar with more ubiquitous narratives about the crisis of the State of Israel's own existence and continuity, privileging certain kinds of victimhood over others.

Two committees to legislate ovum donation: the common goals of the secular left and the religious right

In the aftermath of the emergence of this story, several branches of the Israeli government intervened to rework the laws concerning egg donation. The media began to report about a law proposed by the Committee on the Status of Women. It emerged subsequently that the Minister of Health, objecting to the proposals of this committee, set up his own committee (the Public–Professional Committee on the Matter of Egg Donation).[11] Both Committees were, in a sense, dedicated to facilitating the reproduction of 'infertile' women and were using the language of the promotion of women's 'rights' to reproduce.

The latter committee, that set up by the Minister of Health, was headed by Rabbi Dr Mordechai Halperin, who was both a rabbi and a gynaecologist. Other physicians, rabbis and bioethicists also sat on this committee. Officially, it was set up in order to find ways to remedy the shortage of eggs. However, prior to this, the Committee on the Status of Women had already been debating the issue of egg donation and had themselves proposed a law. This earlier committee was headed by the left-of-centre Member of Knesset (MK) Yael Dayan and Ahmed Tibi, MK, a Palestinian Israeli politician and gynaecologist. They had attempted to table this law in the Knesset before, but had come up against intense resistance from the then Minister of Health (who was a member of the Orthodox Jewish Shas Party). Their proposed law was written with the assistance of patient, advocate and egg recipient, Ofra Balaban (who has an MA in Policy Research).[12]

The fact that there were two separate government initiatives to legislate ova donation indicates that perhaps the 'crisis' touched on issues of national significance. The aim of the new legislation was to expand the egg donor pool in Israel and to avoid further egg theft.[13] Yet the Minister of Health rejected Dayan and Tibi's proposed law because this committee refused to agree to the setting up of a 'donor offspring registry', as part of the new procedures for liberalized egg donation. This would involve keeping a computerized list of donor offspring that would be accessible to religious and medical authorities. Furthermore, donor offspring would be able to access this registry once they reached the age of 18. This secured database would bar all others from access to information contained therein.

The religious authorities wanted the registry in order to prevent the marriage of genetically related siblings. Dayan and Tibi's committee argued that this registry should not be kept under the auspices of the national rabbinate. They were anxious about the intrusion of religious authorities into the private lives of citizens. Hence the Minister of Health appointed his own committee.

The Ministry of Health's 'Public–Professional Committee on the Matter of Egg Donation', or Rabbi Dr Halperin's committee for short, tabled the law after 16 long meetings.[14] Their law proposed to expand the donor pool by allowing any woman (and not solely IVF patients) to donate eggs. Donors would be minimally compensated for undergoing treatment. Cross-religion donation would be forbidden. A donor offspring registry would be created and kept under the authority of the Ministry of Justice.[15] Like the meetings of the Dayan–Tibi committee, this committee addressed ethical and moral issues, including donors' health, payment for eggs, as well as other religious and social issues. Halperin's committee also decided to hire a company called InfoFORT to develop a secure database that would house the donor registry. The outcome of all of this decade-long process was the Egg Donation Law that came into force in 2010, allowing all women to 'donate' eggs for compensation (approximately NIS10,000 or about US$2450). They would not be allowed to donate to women of other religions.

The crises of the egg shortage and of the needs of intended ova recipients were thus offered a political solution. This solution was in itself complicated, revealing tensions between religious, medical, secular and left- and right-wing politics in Israel. Yet it may be the case that the rift between the two committees was not as great as the media represented it to be.

Nitza Berkovitch provides a potent analysis of how aspects of women's 'private' lives are strategically enrolled into national struggles through discourses of gender equality. Feminism is used in Israel to mobilize women for national goals (Berkovitch, 1997). The proposed Egg Donation Laws described above were narrated by their proposers, research participants and the media as contributing to women's rights to have children. These discussions were framed within a discourse of women's rights to reproduce. The state's support for such a discourse, exemplified through the establishment of two committees, suggests that the generalized right of women to reproduce is of great national importance. The conflict between the two committees may be seen, in part, as a struggle over the hegemony of secular or religious interests in Israel.

However, even more potent than the divisions between these committees was their *joint commitment* to one goal, which is the continuity of Jews in the State of Israel.[16] Both committees demonstrated a concerted effort to cut the waiting time for ova and to increase the availability of donor eggs.[17]

Transnational import of ova: 'contaminated' and desirable ova

Impatient because of the delays of legislation described above, some physicians joined forces with Israeli business associates abroad to set up egg donation clinics outside of Israel. One physician set up a clinic in Romania (Chapters 2 and 3). The procedure at his clinic is such that sperm from Israel are brought to Romania where they are combined with ova from paid donors, frozen, then brought back to Israel. This part of the story (of doctors bringing ova from abroad) resulted in another 'crisis' for Israeli ova donation.

After the aforementioned egg 'import' had been practised for a few years, an Israeli newspaper published a statement by a rabbi doctor who argued that the eggs from Romania might be dangerous because they arrive from a country that has a high rate of 'AIDS' (Freilin, 2000). The story was covered in the various national newspapers, on the radio and on television. Almost immediately afterwards, the Ministry of Health issued a restriction on importing eggs from Romania. These eggs were seen as 'tainted' and initially a panic ran through the community of intended egg recipients, physicians, lawmakers and the media. Romania was framed as a potential source of 'contamination' and pollution of the national body.

However, after the initial crisis about contaminated ova occurred, many intended ova recipients were told (by their physicians) of medical evidence demonstrating that oocytes do not carry the HIV virus. Hence, many intended recipients and clinics purveying the eggs characterized the Romanian donors as a desirable source of eggs because of their Europeanness, level of education (donors were said to be university students) and youth (see Chapter 4). Like the semi-permeability of borders of the State of Israel to immigrants of different nationalities, borders were similarly policed in relation to the movement of 'foreign' eggs (but Chapter 4 discusses the comparative desirability of these eggs to ones from Arab women or non-white women).

A court ruling: egg recipients' 'need' as an (national) 'emergency'

When the importing of ova was banned, several prospective recipient women filed a petition to the High Court of Justice to appeal the ban. Their lawyers argued on their behalf for the 'return' of 'their embryos' from Romania. These women's partners' sperm had been taken to Romania to be inserted into donor eggs. It was intended that the donor eggs be brought to Israel, having been fertilized with Israeli men's sperm.[18] Despite the ban, the women's lawyers argued, the fertilized ova belonged to the Israeli couples. After a brief court case, the fertilized ova were deemed to belong to the commissioning couples.

One of the lawyers representing the 90 petitioning women described the case in this manner:

> The problem [of a shortage of eggs] exists, and we had to legally answer it and there was a disparity between the legislator[19] and the Ministry of Health which on the one hand said yes, then said no [...]. It was very quick [...] [a] 45-day process [...]. It was treated as an emergency because of the age of the women [...]. (Daniela Reyzl, lawyer, interview, Israel, 18 September 2002)

Reyzl couched the case in a language of emergency. This emergency was due to the advanced age of the intended mothers (as stated earlier, women requiring egg donation are often over the age of 40). She posits that the case involved a struggle on behalf of the intended recipients between the government, which had permitted the importation of ova from Romania, and the High Court of Justice, which had deemed it necessary to issue the ban. This focus naturalizes the idea that women (many of whom are over the age of 40) can and should be provided with ova donation services. The idea of the emergency of the egg recipients resonates (but is clearly not parallel) with the emergency of the nation. Both emergencies require various naturalizations.

Reyzl indicates that while the Minister of Health was against the import of eggs from non-Jewish women[20] (despite the rabbinical approval of such a transfer), the courts oscillated between permitting and restricting the import of eggs from non-Jewish Romanian women. Whilst the courts initially permitted the import of ova, the HIV 'crisis' led the court to rule against the import of ova, and finally after Reyzl's team and the prospective ova recipients put forth their case, the court once again permitted the import of eggs.

Reyzl seems to be arguing that the 'emergency' of the ova recipients arose not only from their advanced age, but also from the discrepancies between the religious and secular factions in the government. Yet, her hint at this secular–religious split is problematic. That is, as I began to articulate earlier (and as others have demonstrated better than I), perhaps the 'secular' and 'religious' factions actually share a common goal of creating the nation as Jewish, rather than being at odds with one another. It may be that articulating the split between secular and religious factions actually obscures the interdependence of secular and religious concerns that, as Guy Ben-Porat (2000: 223) has argued, consolidate Israel's exclusionary state practices.[21]

Not long after the ova recipients filed their petition for the return of their embryos, the High Court of Justice changed its position again and permitted the entry to Israel of fertilized ova from Romania. The following is an extract from the Israeli Hebrew-language newspaper *Yediot Aharonot* describing this shift:

> **Approved, Import of Ova from *Hul*[22]**
>
> Good news for women who are in need of egg donation, and can't attain them here in Israel: last week the Minister of Health signed an amendment that permits the importation of ova from 'hul' and their implantation in women in Israel.

> An announcement of this was passed yesterday to the High Court of Justice following a plea that was submitted by couples who have no children who claimed that the Ministry of Health restricted, in a harsh manner, the import of eggs. Among the pleas, which were presented by lawyer Yehoshua Reyzl [husband and legal partner of lawyer Daniela Reyzl]: couples asking to give birth to a child by means of implantation of eggs that were donated by a healthy woman and were fertilized in 'hul' from the sperm of the husbands.
>
> The amendments will come into effect only in February 2002, in order for the Ministry of Health to send first a professional delegation to Romania to inspect the clinic, which is where the majority of the fertilizations of the donated eggs are coming from. (Goldberg, 2001)

This article signals the shift, which the excerpt from my interview with Reyzl above describes, in the court's and the government's stance regarding acceptance of 'foreign' ova in order to assist women to reproduce. In this newspaper article, the imagined danger posed to the body and the nation by eggs with 'AIDS' are discussed in a way that brings women back into the centre of the story. While reference to the embodied suffering of the woman (referred to at the beginning of this chapter) was left out of media accounts (after the initial reports) in favour of discussions of how to get more donor eggs, here the 'need' to reproduce was taken to be an emergency that required a solution. In order to describe this emergency, the plight of Israeli women was invoked. The attention paid to ova donation by the courts and the media appears somewhat excessive considering that only a small percentage of the population actually seeks ova donation. Yet this was framed as a social imperative, and reinforces the connection between the 'one and the many' in Israeli nationalism.

There were several different discourses and ideologies simultaneously at work here. First, the discourse of equality and equal rights for all people was being mobilized.[23] This discourse functioned within an ideology of neoliberalism, which presumed equality of all people in Israeli society. The discourse of women's rights to have children intersects here with the discourses of migration, contamination and the Jewishness of the state. The ova were barred from being 'returned' to Israel because of fears of contamination, danger and foreignness (Freilin, 2000). The significance of the idea of 'events' or 'episodes' to anthropology can be examined in order to help develop an understanding of the implications of crisis to Israeli nationalism.

Unravelling the stories

There is a need to think through who the 'victim' is in stories of egg donation and Israeli nationalism and what the act of claiming 'victimhood' means in these contexts. The way in which the crises in egg donation were described to me by clinicians, lawyers, recipients, Ministry of Health officials and in the media as affecting egg *recipients*, rather than *donors*, is significant. What seemed to be given primary attention in the accounts was the 'desperation' of infertile (Jewish) Israelis seeking to have 'continuity', to reproduce themselves. This is significant, given that motherhood in Israel is seen as a 'national mission' (Berkovitch, 1997: 605), where women are 'bearers of the collective'.

Thinking back to the violent everyday political situation in Israel and Palestine during the time when this research was conducted, it is possible that the banality of the idea of a 'war over home' obscures other kinds of violence that were enacted by the state. While privileging the victimization of Israelis who were being killed by suicide bombings, the national media obscured the violence endured by Palestinians. The story of the egg donation crises includes a similar confusion over the role of the victim. The woman, whose embodied experiences precipitated the 'crisis', whose ova were stolen and whose life was threatened by ovarian hyper-stimulation syndrome, disappeared in the many versions of the story. Like an ovum being extracted and moved elsewhere, the focus of the story was wrenched away from the body of this individual woman.

The newspaper and interview accounts that I collected about the crises tended to focus not on the woman herself, but rather on the 'theft of eggs' that led to a 'shortage'. Her story, like her ova, had already been 'exteriorized'[24] and became part of a bigger national story about eggs, theft, shortage and crisis. Tension over the 'war to make a baby' and the emergency of the need of ova recipients overrides the suffering of this woman. The need of recipients also overrides the aspirations of Romanian ova donors discussed in Chapter 2. This tension between the differently positioned women in the stories about the crises in Israeli egg donation helps articulate the relationship between the scale of the nation and the scale of the body.

The ova recipients and the donor whose ova were stolen are figures in the Israeli accounts of egg donation crises. They are discursively brought into being, through repetition of the story, as particular kinds of beings that materialize and represent aspects of a national crisis. The figures of the recipients in need and the woman whose eggs were stolen can be examined in terms of how women's bodies reproduce nations.

Resonances

These media accounts began with reference to a woman who has been 'victimized' (had her ova stolen and became ill). They then move onto extensive descriptions of the victimizers, two male doctors. Subsequently, emergent stories are about the shortage of eggs. This is followed by two kinds of solutions, one governmental solution in the form of a law, the other a 'medical' solution in the form of doctors bringing ova from abroad. Yet, another crisis is said to develop at this point, because there is tension about what *kind* of eggs can be brought into Israel. References to the religion of donors and the potential contamination of ova are brought into the story. Finally, the media reported that the High Court of Justice had approved the import of ova from this agreed source, juridically superseding the cultural fear of pollution. Yet in contradiction to this acceptance of cross-border cross-religion donation, the new Egg Donation Law of 2010 prohibits cross-religioun donation within Israel. So the borders of acceptability of 'forcign' ova are semi-porous and always shifting.

This series of stories, when put together, makes one larger story, which if we take a step back and picture the shifts and foci of the stories provide a different kind of story.

Same, same, but different

The account of the ova donation crisis described above shares some aspects of its rhetoric with more ubiquitous narratives about the crisis of Israel's own existence and continuity. The narrative about crisis *is* a (micro-)Zionist history of Israel. Yet it may be that I have constructed it as such because of the context in which I conducted the research. This egg donation story might have been different had the researcher been a different one, had the Second Intifada not been at its height. Again, we are here in the arena of ethnographic productions, and I want to stress that this kind of production should be contextualized within the broader events taking place at the time.

Moving from crisis of the victims at the hands of a perpetrator, to the establishment of a state, to the debates about demography and the mass migration of people from other countries (some Jews being seen as appropriate, some being seen as potentially polluting to the national body), the mainstream accounts of Israeli state history depict a sense of constant 'crisis'. This is an aspect of 'banal nationalism' in Israel (Billig, 1995). The newspapers and television 'flag the nation' daily. Yet unlike Billig's account, which argues that banal nationalism differs from extreme nationalism, it seems that everyday crisis and violence *is* the banal in Israel.

Back to the arena of ova donation, the banality of crisis was evident in my own initial reception of the stories of the 'episode that exploded'. Initially, this 'episode' was an interesting story that indicated to me that some aspect of the 'local' in Israel was provoking intense public debate about egg donation. Based on this I decided to examine everyday discursive practices of egg donation, trying to see the 'tacit' meanings of these practices. In the early days of fieldwork I attempted to move my interviewees *away* from telling me the story of the 'crisis' and yet, almost every person whom I interviewed began with telling me about 'the crisis that had befallen egg donation'. As a fairly new ethnographer at the time I initially ignored these and tried to obtain a different story. Yet it is extremely salient that nearly every person involved in egg donation narrated it as being 'in crisis'.[25]

The way in which these crises were narrated has resonances with the national crises about the nation's right to exist as a Jewish state, and about being in a 'war over home'.[26] The way in which Israeli crises were narrated to me as placing the State of Israel and the lives of Jews at risk seemed in some ways to rely on the idea of a permanent 'state of emergency' in which Jews had lived in Europe. That is, in the daily narrations about crisis there resonated elements of the national discourse of the violence suffered by and victimization of Jews, which had taken place in Europe under the Nazis. The very establishment of the Jewish state has often been represented as a remedy for the deaths of Jews at the hands of the Nazis (Almog, 2000).

This persistence of violence and victimhood is uncanny in that the early Israelis modelled themselves in direct opposition to the weak, victim Jews of Europe, self-identifying instead as a nation with military prowess (Almog, 2000; Weiss, 2002). Oz Almog traces this self-modelling of early Zionist colonizers of Palestine through the trope of the '*Sabra*', which was a key symbol for those early Zionists (Almog, 2000). The term *Sabra* comes from the *sabra* cactus fruit, a prickly pear. It is an idealized self-representation that the state hoped that Jews would take on in Israel. This trope was held in contrast to tropes of the 'old' and 'damaged' Jew of Europe. The 'new' Jew, or *Sabra*, was represented by the first generation of Jews born in Israel, representing all that was 'better' about them. Ironically, the sabra fruit was planted by Palestinian farmers and used to mark the boundaries of landholders' fields. The new Jews were imagined as strong of body, hardened fighters on the outside and sweet on the inside (like the *sabra* fruit). Indeed, it is a very Europeanized 'new Jew' that the *Sabra* represents (Weiss, 2002).

It may be the case that the accounts in which I read and heard about the crises of Israeli egg donation were in some ways about a tension regarding the place of 'victimhood' in Israeli society. While on the one hand the position of the victim is vied for, it is on the other hand also a feminized position that negates the necessary strength, militarism and masculinity of idealized Israeliness. The continuing thread is from violence experienced to violence perpetrated and re-experienced.

Ova recipients: oppressed figures as national heroines

For Walter Benjamin (1999 [1947]), 'emergency' was identified as a constant state in which historically the oppressed live. In some sense, Jews and non-Jews in Israel are still living in a state of emergency perpetuated by the government's use of them as a military front-line. The constant, everyday-ness of emergency, violence, crises and explosions in Israel, the banality of crisis, is what upholds the State of Israel. This everydayness of crisis and emergency obscures the 'artificiality' of the state's own inception. That is, the origin myth of the state hides the fact that it was not solely realized through victimhood but also through forceful violence of extraction. The uncanny present-day victims of the Shoah cannot but reproduce that violence and still feel that they are oppressed by it.

The banal nationalism associated with the narrative of crisis seems to figure also in the accounts of the crisis of egg donation. Tension over the 'artificiality' of the motherhood of ova recipients can be overridden by accounts of their crisis, and by figuring recipients as 'victims'. The fact that the crisis of the egg theft was referred to as 'the episode that exploded' is one instance in that narration that most closely ties together the explosiveness of everyday life in Israel with the national significance that I have suggested is present in the story of the crisis of egg donation.

The role of ova recipients as figures in the account of the crisis is complicated. While the *individual* woman at the start of the account (whose ova were 'stolen' and who became ill as a result) disappears from later newspaper reports of the shortage, a central figure is the *collective* group of women in need of eggs. An individual's crisis was made into and taken up to support a claim of a crisis of a collective. This collective was characterized by their lawyer as being in a state of 'emergency' due to their age. In this sense, their normalization as Israeli mothers requires their pathologization as 'too old' and 'abnormal'. As victims, they can become national heroines.

As Nira Yuval-Davis (1997) has argued, women are the 'border guards' of the collective. Yet ova recipients may collectively be 'a figure' in a narrative that upholds the nation. In a society that valorizes motherhood to the point of referring to women who have 12 children as 'heroine mothers' (Yuval-Davis, 1997: 30), and which at the same time relies on a discourse of victimhood to legitimate its existence, the ova recipient may be a powerful figure, which, as Claudia Castañeda has stated, 'not only condenses material-semiotic practices, but also brings a particular version of the world into being' (Castañeda, 2002: 4). This figure of the ova recipient may be one more site through which the nation is narrated as having a precarious existence. The media accounts and women's own accounts of the recipient's need and state of emergency may also be the sites in which this precariousness is figured. What can be seen as a pro-women or even feminist centring of 'the female subject' in newspaper articles, legal arguments and committee reports may also be a way of performing nationalism through the figure of the ova recipient. Narratives or ideas of crisis and women's bodies are brought together as the most available resources in a particular kind of banal nationalism. Reference to the women's needs invokes authenticity, nature, humanness, vulnerability and heterosexual normativity in which the stories of the theft of eggs and egg shortages are situated.

The focus on crisis and victimhood in this chapter is a route for experimentation in ethnographic writing. How to integrate the challenging experience of living, researching and transforming the self as a political actor and citizen (I am thinking of my own transformation during the time that I conducted this research), with a portrayal of what might have been happening with egg donation policy and practice at the time (indeed, there were so many versions of the story that it was difficult to grasp even when in its midst).

The majority of my interviews centred on peoples' experiences of pain, sadness and their desire for a child. They were not about nationalism. Yet nearly every interviewee talked of egg donation as being in crisis. It would be remiss not to make something more of this ethnographically, especially as it confirms my own shifts in consciousness whilst in 'the field' and afterwards. I could have argued that the suffering of Jews at the hands of enemies produce us as victims throughout time. This is a comfortable and familiar narrative. Instead I have highlighted the ways in which these narratives participate in a whole series of claims to crisis of the Jewish state that legitimate that state's existence, despite the fact that the state is very much the aggressor in much of its brief history.

7
Synecdoche: Writing a Transnational Egg Donation Story

I began with the suggestion that egg procurement practices make nationalism palpable. But this can also be said in reverse, that nationalism makes egg procurement practices knowable. Romanian egg sellers may be exploited subjects, and the practices of oocyte extraction at the private clinic in Romania did not always manifest the 'care' dimension of 'cross-border reproductive care'. The practices at this clinic indicate a 'reverse traffic', a way in which the liberal democratic policies of the European Union for protecting people against trafficking can be bypassed by reversing the direction of travel and changing who travels. In this case the eggs go to the recipients and the doctors travel instead of patients. The Romanian women whom I interviewed saw themselves as benefiting and even profiting from their situation, they transgressed Romanian paternalism and tried to 'get a better life' at the same time as they were exploited for their valued body part. The doctors see the Romanian women as an available resource, which can help them make good reproductive citizens of 'un-reproductive' Israelis. In this manner the private clinic becomes more of a supportive family to egg recipients than the state. This is done through the cultivation of Jewish notions of messianic hope, coupled with capitalist ideals of market exchange and profit. Women receiving eggs are seen as wishing that they did not have to travel for eggs, wishing that they did not need eggs at all. Indeed they wish to be invisible. It is a lengthy process to accept 'this thing' of getting an egg from another woman. In the process, they bond with one another, gaining an almost military sense of solidarity. They are encouraged by the clinics to think carefully, selectively, about their donor/egg/baby in terms of desirable and undesirable traits. Some women reproduce notions of European whiteness in their selections. Meanwhile, minoritized Mizrahi women occasionally resist the imposed borders of Jew versus Arab in their imaginings of who is an acceptable donor. In so doing, they remake the border between Jew and Arab.

These borders however are carefully, sometimes violently, guarded, in both egg donation practices and the everyday world in which those practices occur. The violence of the metaphors of oocyte extraction (*dikkur*), Assisted Zona Hatching (AZH), Intra-Cytoplasmic Sperm Injection (ICSI), betray the fear that Israeli people live with daily as their state 'disposes' of them, as one doctor joked. This kind of joking, so prevalent in Jewish Israeli life, betrays the fear that the state that should be protecting the people puts them more in harm's way. This is further exemplified in the 'episode that exploded' in the Israeli and global media about the theft of oocytes. This episode, the other crises that befell egg donation and the extracts from my fieldnotes about the everyday crisis in Israel, describe the quotidian sense of crisis with which Israelis live. Through repeated crises around egg donation, ova recipients get made into national heroines, whilst the unknown donor whose eggs are stolen fades into obscurity. This is a banal crisis of nationalism as much as it is the banal crisis in capitalism.

Extractions as exceptions and 'universals'

The notion of 'local biologies' has been extremely helpful to feminist anthropologists aiming to critique the universality of women's biology (Lock, 1995). Undermining the idea of a universal female biology helps critique the way in which women are made subordinate in different contexts. This has been an important feminist anthropological intervention into the construction of gender, nature and biology. In the Israeli case, however, a local biology can be taken up in state rhetoric to denote the special nature of the Israeli state (as in examples that foreground the protection of Jewish lives over Palestinians'). A major claim about Israeli reproduction has been that it is unique and local because of Jewish rabbinical thinking around kinship, Jewish historical experiences of violence and the present-day crisis in which Jewish Israelis find themselves surrounded by enemies. Seeing only the 'local' here would prevent a deeper sense for how biology, nature, history and capitalism are enacted in Israel and in reproduction more broadly; the mutual materialization of nationalisms and global markets. Or which lives are more worthy. A comparison of two national contexts might equally harden 'the local'. *Extractions* keeps the particular central (through detailed attention to the everyday things that people do and say), whilst also attending to the more universal dimensions (unpacking broader ideologies). This foregrounds a central tension in the social sciences between hermeneutic interpretation and ideology critique (Lambek, 2011). Israeli transnational oocyte procurement, fertilization and transfer practices confirm that reproducing Jews is both universal and particular. Indeed the particular tells a story about the universal and vice versa. The 'local' and 'global' become nonsensical concepts separating as they do things that are always interlinked (Das and Poole, 2007, cited in Lambek, 2011).

Assisted reproductive technologies brought forth a change to Euro-American kinship. The act of conception, need for a blood bond and the numbers of parents all shifted. New ways of conceiving of the natural body and biology were required in order to understand how families were made. Meanwhile, Jewish rabbinical laws dictated that a Jewish person was made from a Jewish mother. This is a matrilineal society, in classical kinship-speak. An egg taken from another woman should unsettle this kinship cosmology. It should confuse it. However, after much debate and disagreement rabbis decided that the mother would be the person gestating (carrying) the fetus, not the woman who gave her egg (Bick, 1997; Bleich, 1997). So, gestation not genetics made the kinship bond. This meant that Jewish notions of matrilineality were not unsettled by reproductive technologies, but that Jewish religious thinking and the contemporary need to protect Jewish lives resulted in maintaining a uniquely Jewish kinship cosmology, distinct from the Euro-American one (Kahn, 2000; Prainsack and Hashiloni-Dolev, 2009).

For a period of time the rabbinical notion of matrilineality held sway in policy dictates around motherhood. Around the time of the Second Intifada, discursive practices around Israeli borders and kinship changed in the Israeli social and political arena. Everyday violence enacted and feared by Israelis led to a resurgence of policies and technologies of removal of the Arab 'Other', through building fences/walls, reinstating demographic policies, suggesting the removal of Arabs from the state. This led in more recent years to laws demanding declaration of loyalty to the Jewish state, policies of deporting non-Arab foreigners, and now it is being suggested in the Knesset that mention of Nazism in comparison to the Israeli state be an offense bearing a fine of US$25,000. The Egg Donation Law prohibiting cross-religious egg donation is deeply entangled with this hardening of borders and kinship. Now eggs too became the bearers of something other than just the promise of a child for Israelis. They were (as they had been for Israeli people throughout the course of my research) something that bears and confers deeper national, racial and historical meanings. This new kinship and border thinking resembles Euro-American kinship thinking (which was never spatially bound to Europe or America in Strathern's account, 1995).

Israeli lawmakers, doctors, scientists and prospective parents are highly influenced by European laws, conventions and ideals. It should come as no surprise then that European ideals of relatedness should travel as well. A Jewish mother may result from getting another Jewish woman's egg. But a Jewish mother and a Jewish baby can no longer result from a 'non-Jewish' egg. Rumours spread of cross-racial breeding prompted an Israeli company to develop a product ('embryo guard') that would monitor the incubators into which embryos were placed in order to ensure that no racial mix-ups would occur. This is European blood-relationship kinship and racial thinking to the core, mixed together with wider concerns and technologies of securitization. These are the repro-migrations or the movement of reproductive ideals and materials that make states, nations and bodies.

Israelis are profoundly supportive of the use of technology (Prainsack and Hashiloni-Dolev, 2009), whether they be religious or secular Jews. Non-Jewish Israelis too have become willing users of technology in a bid to be 'modern' (Kanaaneh, 2002). This modern attachment of Israelis to technology, however, may confirm something *other* than the uniqueness of Jewish kinship thinking. It may in some sense show precisely the opposite. A desire to be technologically advanced, as well as leading the world in global economies of tissue exchanges and preferring traits that confirm whiteness, all demonstrate the very Western or Euro-American nature of kinship ideals in Israel.

That the Israeli state competes so well in global techno-markets and bioeconomies shows how much of an emblem of the wider global capitalist markets and racial histories of migration this state is. The interconnectedness of 'local' affect, ethics and histories with the rest of the Western world is demonstrated through different modes of extraction. Arguing the local biology and kinship in Israel is what the Israeli government does daily. An important ethical intervention is thinking about how as anthropologists we reconcile the particular claims of the state and daily practices and discourses of people with more universal discursive practices.

The practices in *Extractions* show the ways in which the focus on the particular to the exclusion of the universal in this context is a profoundly ideological act. Israeli kinship, nature and biology are entangled with capitalism, histories of imperialism and migration. Jewish notions of matrilineality formed in part during attacks on Jews *as* Jews. Contemporary Israeli kinship thinking (including race, religion and nationality) is formed in the context of *rhetorics of the state about* attacks on Jews *as* Jews. In this context it would perhaps be better to think of Israeli kinship rather than Jewish kinship. However, it is also an ideological act to focus on both. 'Jewish experiences in Israel are unique' is a nationalist claim. Conversely, 'Jewish experiences of their state disposing of them as part of international webs of markets, ideals of bodies and protection of wealthy states' interests' is also an ideological statement (this book was always already an intervention into thinking about global ethics).

Making interferences

There has been a tension in writing this book between anthropology as descriptive analysis and anthropology (and feminist STS) as critique. Whilst attempting to be faithful to the accounts of the doctors, nurses and embryologists who gave so generously of their time, at the same time I came to many conclusions with which they would probably disagree. As an 'insider-outsider', an Israeli who has lived outside Israel for the majority of my life, I have the liberty to critique the state without repercussions that my Israeli colleagues have. It has been both a feminist urge to analyze what happens to gender when it gets mixed with nationalism, and an anthropological urge to see what happens to nationalism when it gets mixed with gender (which it always is of course). Anthropology is experiencing a push both from within and without to be a 'public' discipline that intervenes in culture (Nahman, 2012). Some have made very strong arguments for the ethical imperative to play a 'public' role, suggesting that human liberation be a central aim. Anthropology, 'has an opportunity to become an arbiter of emancipatory change not just within the discipline, but for humanity itself' (Scheper-Hughes, 2009: 4).

I employ Donna Haraway's (1991) notion of cyborg feminisms to enable critique through feminist writing that accounts for the fictions that one is creating (because writing culture is always a kind of fiction – although not necessarily ficitonalized; see Visweswaran, 1994; Geertz, 1988; Clifford, 1986; Handler and Segal, 1990; Strathern, 1991). One way of doing cyborg feminism is through 'diffractions', discussed in the Introduction as a technique of making new stories by filtering them through a particular eye/I. Each chapter has taken one particular 'site' of the making of egg donation and the nation and attempted to engage with a problem or set of problems raised for feminism or nationalism. In Chapter 2, this was the feminist debate about exploitation and trafficking, answered through a detailed walking through the clinic and the sights and voices within. In Chapter 3, this was about how the private and the national come together transnationally, seen through logos and company mottos. In Chapter 4, this was done through an emphasis on becomings, intimacies and repro-migrant subjectivities. In Chapter 5, deep description of the socio-technical practices in the lab, clinic and surgery theatres was intersected and interrupted by violence in streets, violent metaphors and the fears of Israelis. Chapter 6 builds on all of the quotidian examples of crisis from the research, mixes it with the crises narrated to me about egg donation in order to suggest that Israeli banal nationalism is one of crisis, and that egg donors and recipients are locked in a battle of sorts over the role of 'victim', which is ultimately the role taken up by egg recipients in the public narrative of these crises. This battle is synecdochical with the national battle for the status of victim.

Materialized through the hopes of donors for transgressing their present poverty, eggs become a mode of accumulation; an alienated body bit whose value is its ability to help all to profit, though some less than others. Imbued with hope, they provide recipients with the seed for the possible beginnings of a family and of hope and personal salvation as good citizens. Doctors, nurses and embryologists see themselves as enabling this achievement of personal and national goals. Yet in this context of the State of Israel, making and maintaining the state is done through violence, crisis and victimhood. This bears echoes in egg donation practices, be they procurement, removal, fertilization or transfer into the wombs of women. What all this does is hint at the way in which exploitation, economics, politics, gender and love work in the broader world in which we live in.

Ha-Mifletset, 'The Monster'

Set against the backdrop of Jerusalem pine trees, *Ha-Mifletset* ('The Monster') (Figure 7.1) is a condensed signifier of femininity and impurity in Israel. Across the street from her and pointing in her direction is a road sign saying 'To Bethlehem'. *Ha-Mifletset* is a good vector for carrying the stories that I have been telling. They slide down her forked tongue and onto the page. I used to pass this huge sculpture daily during my research in Israel. It is as much a part of the field as the clinic, surgery room and lab. *Ha-Mifletset* reminds me of the monstrous social situation in which this research was carried out, of the structured but somehow unruly nature of IVF across national borders, in public as well as private medicine. Finally, it reminds me of the cobbling together of theoretical approaches to make a coherent narrative about egg donation.

Commissioned by the city of Jerusalem in 1972, the outdoor sculpture commonly referred to as *Ha-Mifletset, The Monster,* was created by Nikki de Saint Phalle, a French surrealist artist. Officially it is called the *Golem,* but everyone knows it as *Ha-Mifletset.* Origin stories of *Ha-Mifletset* have gained the status of urban legends. Her incarnations multiply. When my aunt came to visit me in my Jerusalem flat she passed *Ha-Mifletset* and recalled that she had once been on a tour of Jerusalem sites and that a tour guide had told her that this sculpture was made by a French artist who had intended it to represent madness and the uterus (a theme that fits well with Saint Phalle's other work, which rejects standard ideas about womanhood and expectations of motherhood). Indeed, *Ha-Mifletset* looks mad and her uterine horns unruly. My aunt said that the city could not refer to it by the artist's original intent because of — '*kefiyya ha-datit*' (the religious coercion) that images in public spaces must obey the principles of 'modesty'.

Figure 7.1 *Ha-Mifletset,* 'The Monster'
Source: The author.

The images of the monstrous and the Golem both do something for this egg donation story. The Golem is a mystical Jewish figure, and one was made out of clay in Prague in the sixteenth century. This creature was supposed to protect the Jews from attack, which the community was tragically regularly subject to. The Golem sadly became uncontrollably violent and began attacking and killing people around it, and in some of the myths he even killed and attacked Jews instead of protecting them.

Many, many Israelis today are unhappy with their government and feel that it is creating a state where Jews are more vulnerable to attack than they are elsewhere. I have been writing this book during a time of the resurgence of mass protest in Israel against neoliberalism and the decline of the welfare state. That movement known popularly as J14 involved tent cities spreading across the country and protest marches of hundreds of thousands. In the summer of 2011, it seemed that Israelis were waking up and demanding mass change. One issue, however, carefully remained outside the frame of this protest. This was the issue of the occupation. Anytime that people or groups attempted to bring this into the debate and discussion, to demonstrate broader entanglements of global markets, borders and racism, they were told to be quiet and focus on the issue at hand. This was a disappointing moment for many Israelis on the left who had hoped that this would be the day on which the Jewish value of '*tikkun 'olam*', 'healing the world', could truly be realized in Israel.

This book is a kind of feminist monster, in the two senses of monstrous that are to show, 'monstrare', and to warn, 'monere'. Borrowing and suturing together anthropology, feminist technoscience studies and postcolonial theory, at times it is unruly. It seemed to me in writing this that anthropological theories of reproduction and the body could be enabled and assisted by the feminist STS literature. The critique of Israeli reproduction could not have done without the Israeli postcolonial feminist literature that has come up against the minoritization and Othering of some Israelis in ways that perhaps mainstream anthropology of reproduction and STS might not have done. Concepts have been created in the spirit of Haraway's injunction to create feminist fictions that tell a 'truth'. Repro-migrants, extraction, reverse traffic, were used where the literature was perhaps not able to capture both the specificity of what I observed in my fieldwork and its generalizability. It is also hopeful that these disciplines together can do what one of them alone could not, which is to interrogate deeply how power and injustice have worked in Israeli society and in global reproduction. Bringing together these various disciplines can help to demonstrate, on the one hand how matter and meaning are entangled (Barad, 2007), but also on the other hand how to enact resistance to the narrowing down of scope and possibility for a really politically and publically engaged scholarship. Some have called this a 'militant anthropology' (Scheper-Hughes, 2009).

That women's bodies are sourced for eggs and that the State of Israel occupies Palestinians, are brutal but not shocking truths. Rather, they are emblematic of banal capitalism. The jolts and ricochets in the text, or where the seams fall apart slightly, can and should be read as resistance to the way in which histories and disciplines formulate clean narratives that wipe out critique and creativity.

Notes

1 Extractions

1. In this book I refer often to 'Palestinians'. In general, it is accepted that people identifying as 'Palestinian' living in the Occupied Territories of the West Bank and Gaza are referred to as such. In general, people who claim a Palestinian identity but who live within the internationally recognized borders of the State of Israel are referred to by the state (and often by Jewish Israelis and themselves) as 'Israeli Arabs'. This is a contested term, which claims possession by the Jewish state of 'its' Arab citizens. Such a claim allows the question of the historical and present-day occupation of the entirety of 'Palestine' (i.e. the current State of Israel) to go unanswered. The term 'Israeli Arab' blurs the fact that some Jews are also Arabs, and hence can also be called 'Israeli Arabs' (I generally refer to them as 'Arab Jews'). I have therefore chosen to refer to non-Jewish Arabs living within the State of Israel as 'Palestinian citizens of Israel' and for brevity have shortened it to 'Palestinians' in the chapters. There are many slippages and inconsistencies in this practice (e.g. there are Palestinians living within Israel's borders who are unrecognized by the state and are therefore not actually citizens), but to my mind it is the most politically accurate nomenclature.
2. In a televized speech in March 2002, Ariel Sharon, then Prime Minister of Israel, declared this war over 'home' and his words were repeated by many.
3. Alhough this heteronormativity is being challenged and enfolded with Zionism in more recent times (Kuntsman, 2009a; Puar, 2011).
4. Birenbaum-Carmeli and Carmeli (2002), Inhorn (2003), Ivry (1999, 2009), Kahn (2000), Landau (1998), Sperling (2010), Teman (2003, 2010) and Weiss (2002), as well as others too numerous to cite here all highlight the pronatalism of the State of Israel.
5. In 2011, the Israeli government barred all public funding for groups that mark the Naqba.

2 TheoristSellers

1. All research participants have been given pseudonyms throughout the book, except where otherwise stated.
2. Another version of this story was published in Nahman (2011).
3. But I am not arguing for paying Romanian women more money (Thompson, 2007; Widdows, 2009).
4. As previously noted, since 2006, Romania no longer permits the selling of eggs, so this chapter captures a moment in history when this was allowed and happened legally.
5. Although I was able to communicate to the ova sellers and clinic staff in my rudimentary Romanian, many cultural idioms and references eluded me

during my research. I have never studied Romanian formally. Rather, I've picked up the language through hours and years of sitting with my mother and her mother and sister, listening to them speak Romanian (my maternal grandparents' mother tongue). In the clinic I spoke mostly English and occasionally spoke some Romanian. Many of the sellers speak English. When I interviewed a woman who did not speak English, generally one of the sellers acted as translator.

6. To get the timing correct for oocyte production according to the schedule of extraction arranged by the clinic, the injections were given at this time in the evening.
7. See Aihwa Ong's apt injunction that feminists recognize and accept that women in different societies may choose to live differently from a Western feminist ethnographer's ideals (Ong, 1988: 90). Other feminists have critiqued the colonialist tendencies of Western feminists, especially in the 1980s (for examples, see Mohanty, 1984; Spivak, 1987). Others (Gupta, 2006) have called for a universal feminist bioethics, but I have critiqued this elsewhere (Nahman, 2008).
8. Hebrew for 'extraction', see Chapter 5 for a discussion of *dikkur*.
9. The phrases 'doing this' or 'making this' or 'this kind of thing' referred to donating eggs. The phrasing came up repeatedly in my interviews and encounters with staff and sellers at the clinic. I was not sure if they referred to egg donation as 'doing this' because it was an abbreviation. I sensed that perhaps the phrasing was, rather, a kind of euphemism that they used because (and some sellers told me that) they were uncomfortable with the idea of giving a part of their bodies to someone else in return for money.
10. Just as the sellers were referred to as 'girls', Romanian men, in this instance, were referred to as 'boys'.
11. This is yet another instance in which ova and children were collapsed in the talk of the people that I interviewed.
12. 'If finally what remains intact is the intention or desire to have a child, then that desire is what the child "reproduces". So in becoming a means to fulfil such a desire, procreation itself ceases to be the crucial "reproductive moment". We might see that moment as instead the acting out of intention or desire' (Strathern, 1995: 355).
13. Donna Haraway borrows from Bowker and Star (1999) the notion of 'torque' to mean, 'where biographies and categories twine in conflicting trajectories there is torque' (Haraway, 2003: 296). When lives and social and political categories come together, they often do so in ways that are oppositional, yet mutually dependent.
14. My very brief gesture towards the feminist anthropology of kinship and reproduction here is too brief. But it is crucial to note that my thinking on this subject has been greatly influenced by Hayden (1995), Strathern (1995), Franklin (1997), Dolgin (1995) and McNeil et al. (1990).
15. Feminist science studies is a sub-discipline of Science and Technology Studies (STS) that takes as central how gender and power affect and are affected by practices and discourses of science and technology.
16. This perspective is strongly influenced by Aihwa Ong's work, *Flexible Citizenship* (1999), which is an analysis of versions of 'the national', considering the movement of migrants and capital across borders.

3 EmbryoMethod

1. Immigrate to Israeli, literally the word means 'to ascend' (see also Dominguez, 1989).
2. The assumption that most of the people whom I spoke to had was that there was a husband involved. There have been difficulties in single women getting egg donation in the past (see Kahn, 2000) but this has been rectified through High Court rulings. However in most cases people refer to 'the couple' and they mean the husband and wife. None of the women whom I spoke to identified themselves as lesbians and none of the clinicians spoke in an inclusive manner about lesbian couples. Normative heterosexuality dominated discourses around egg donation and IVF in all of my encounters.
3. Tal uses the more delicate term *she'iva*, which implies suction rather than *dikkur*, which implies stabbing or poking (Chapter 5).
4. 'Embryos' is a problematic term as they are not quite embryos yet. This is important as it forms part of a legal 'loophole' through which the return of these embryos was enabled in February 2002. The ruling of the High Court to permit the import of ova read 'fertilized ova' and not 'embryos'.
5. 'Reinserted' seems like an odd way of phrasing it, as only the sperm had previously been in the canister. The embryos are being put inside for the first time. Yet by saying that they are 'reinserted', Tal seemed to indicate that she imagines the embryos as just an extension of the sperm that had already been in the canister. This may be part of the way in which the embryos get imagined as already belonging to the commissioning couple.
6. For all of this care, notation and observation, there were still misplaced embryos that I observed in the laboratory.
7. See Chapter 5 for more detail on how this works in Israeli egg donation.
8. See Rabinow (1995) and Gibbon and Novas (2007) for good discussions of 'biosociality' or the ways in which patients are becoming active in managing their own health. My interviews with some of the women at this clinic indicated that they too needed to be active in furthering their reproductive goals at this private clinic. The frustration that they felt at trying to get the *kupot holim* to reimburse their costs, or trying to understand how many embryos they would have implanted or why they had to wait so long is indicative of this contrast between the company ethos and patients' experiences of it at times. *Kupot* (s. *kupa*) in this context is the short form for *Kupat Holim*, or Sick Fund, meaning the national healthcare provider.
9. We have many accounts of the relationship between the Israeli state, nationalism and the body. Many of these point out the militarized nature of Israeli society; whilst others discuss the ways in which religion plays a role in Israeli negotiations of kinship, medicine, genetics. In terms of IVF, the state is typically seen as benevolently assisting the reproduction of citizens for demographic and survival reasons.
10. Mahmood (2005) and Asad (2003).
11. The politics of race with relation to this 'pool' of donors is discussed in Chapter 4 and Nahman (2006).
12. 'There is an uncanny quality to the production of the state through the production of an enemy because often the criminal or terrorist or threatening

Other is a familiar face, familiar but strange, strange in its familiarity' (Aretxaga, 2003: 404).

13. But also see Becker's important work on the slippage between embryo and baby (Becker, 1997; Becker, 2000).
14. The U-turn in Israeli law about who can donate eggs to whom is a clear case in point. It is now illegal for women of different religions to donate oocytes to one another within Israel.
15. Whilst I am completing the book, at present the mass protests happening in Israel against privatization and consumerism indicate the ways in which this 'choice' is limiting, impoverishing and is a mask for more destructive state policies.
16. Kanaaneh (2002) writes about how abortions are selectively targeted at Palestinian women by both the state and sometimes by Palestinians themselves in an effort to be 'modern' and therefore have fewer children. Amir and Benjamin (1997) have written about the way in which abortions are targeted at Mizrahi women. More recently, it has come to light that in communities of Ethiopian Jews in Israel there has been targeted use of Depo-Provera, the implant/injection contraception (Chehata, 2012).
17. The significance of 'victimhood in the context of ova donation and Israeli nationalism is discussed in Chapter 6. See also Kuntsman (2009b).
18. Sander Gilman provides an important analysis of Jewish anti-Semitism in Europe that gives a sense of the racial topography of Jews in the nineteenth and early twentieth centuries (Gilman, 1991). As an Eastern European country, Romania would once have been seen as less desirable; however, probably due to the large immigrant population in Israel from Romania, there has clearly been a shift in this preference.
19. Names of two large dessert- and dairy-producing corporations in Israel.

4 Repro-Migrants

1. The (much contested) need for secrecy and anonymity in IVF and gamete donation is not unique to Israel (Bharadwaj, 2003) and has been well documented in other contexts. As Marcia Inhorn has illustrated in her study of infertility in Egypt, secrecy is often deemed necessary due to the social 'stigma' attached to infertility (particularly male infertility). Drawing on Goffman's (1963, cited in Inhorn, 2003) ideas on the management of stigma, Inhorn argues that IVF has become morally loaded in Egypt and is a 'constraint on the use of the new reproductive technologies in Egypt' (Inhorn, 2003: 263). The stigma present in ova donation in Israel, however, is far from a constraint to its use.

   In a study of English couples who attempted to conceive using donor sperm, Snowden et al. (1983) indicate that secrecy was important to the majority of people (although nearly half the recipients did not wish to maintain secrecy about the donation of gametes). This study indicates that parents often wish to protect their children and the family from being stigmatized and to 'normalize' the situation (ibid.). This view was echoed by several of my interviewees, who suggested that children need to be protected from the cruelty of other children who would potentially call them names for being 'different'.

Erica Haimes (1993), whose research on donor gamete anonymity is also based on the UK, indicates that secrecy is one strategy that parents use in managing the genetic identity of their children. While some parents do not disclose anything about the means of conception, others disclose the fact that the child was conceived through assisted conception techniques and still others prefer to disclose everything to their child (ibid.). Similar to Snowden's findings, Haimes (1990) found that secrecy about gamete donation is a way of protecting the 'anomalous' family and promoting a heteronormative nuclear family life.

However, research about the harmful effects that secrecy of 'genetic paternity' has on donor offspring (Blyth, 1998) has led to shifts in the UK in terms of disclosure of information about donors to donor offspring. Campaigners for the rights of donor offspring argue that knowledge about the inheritance of genetic disease and other crucial psychosocial issues such as 'knowing' the identity of the donor, for example, are important to them. This has led to shifts in policy where individuals in Britain who were conceived using donor sperm can access information regarding their genetic 'parent' at the age of 18.

2. I interviewed 25 ova recipients. Most interviews were conducted at the clinic in Tel Aviv. One interview was conducted at the hospital in Jerusalem. Two interviews were conducted at recipients' homes, two at cafés and one at my house. The duration of interviews was generally between half an hour to an hour, although several interviews lasted for longer than an hour and some for up to three hours. One ova recipient couple, Ofra Balaban and Itamar Kastelansky (their real names), are the heads of a national patient advocacy and support group called CHEN. I interviewed them at their home and visited them several times during my research. I also met with Ofra at events, which she had organized for members of CHEN, and occasionally bumped into her at clinics. Of my 25 interviews with recipients, 14 were with women and their husbands, and 11 were with women on their own. The interviews were semi-structured and tended to begin with the invocation 'tell me about how you arrived at egg donation'. I tried to allow the encounter to be interviewee-led as much as possible. This means that I would ask interviewees to tell me what they thought I should know about Israeli egg-donation and about their own experiences of it. This tended to generate responses that were about personal histories of 'infertility', marriage and family, as well as about legislation and the government's role in facilitating egg donation.
3. Another group of women who often cannot conceive but were not discussed with me in interviews are those who were born with 'ambiguous' sexual organs. In North America at least, such women have often had their ovaries, or ovo-testes, removed at an early age because of cultural anxieties about sexual ambiguity. One woman whom I interviewed in Israel had this done to her as a child. Such a 'condition', which in North America and the UK is termed 'intersex', has wide-reaching implications for the individual in question. For a discussion of the medical construction of sex through the category of intersex and for an in-depth discussion of the variety of experiences of individuals who have been diagnosed with such sexual variation, see Nahman (2000).

4. Mizrahi and Sephardi are often used interchangeably to denote Jews from North African, Arab and other Muslim countries such as Egypt, Libya, Morocco, Iraq, Turkey, Iran and so on. There is a politics to the use of Mizrahi rather than Sephardi. The latter, which means 'Spanish', lends a European veneer to what are largely non-European people.
5. By mixed I mean that often people were neither one nor the other, they could be both. Again, I want to stress that I see identity as a process, that is always incomplete, and so whether or not people identified as Palestinian, Mizrahi or Ashkenazi I recognize the complexity of identity categories, which can never fully describe people (Hall, 1991). People are not Mizrahim – a reified category available for objectification. Rather, Mizrahiyut is a place to spring into social action and criticize Euro-Israeli academic theorizing (Motzafi-Haller, 2004: 181).
6. It tended to be difficult to categorize people into sociological groupings, for they themselves often resisted such categorization. I am interested in how discourses of race, geography and economy play a part in the discursive practices of ova donation. I identify people by name, and I tend to include information about them if it seems relevant to the particular passage. Nonetheless, in this study I set out deliberately to include women of different religious, ethnic and class backgrounds, which I ascertained through their own self-identification. Also, I did not screen out non-Jewish respondents, for example.

   In analyzing the responses to my interview questions I have tried to account for these varied subject positions but also not to over-attribute people's responses to questions about their ethnicity and class. The following discussion of the ways in which ova recipients construct the racialized boundaries of the nation relates to the *'choices'* they made about which *kinds* of donor they wanted and what kind of child they imagined themselves having. That is, the majority of my analysis is not about the identities of the *recipients* per se. Instead, I focus on the kinds of ideologies about appropriate/desirable national subjects that were produced and reproduced through processes of 'choosing' donors and eggs.
7. She began menstruating. In order for the embryo to implant properly, the blood that lines the uterus has to be 'thick'. A menstruating woman cannot, in most cases, sustain a pregnancy.
8. *Vatteran* comes from the word *le-vatter*, to relent or concede. In my experience, *vatteran* in Israel is generally a feminized characteristic that is not a virtue since masculinity is valorized (Weiss, 2002). It is associated with weakness. This term reminds me of the ubiquitous phrase, *'Lo nevatter 'al Ha-Shtahim'*, the right-wing slogan, 'We will not concede the Territories'.
9. This proverb may have multiple origins and certainly the sentiment is found in both Swedish and Japanese cultural traditions as well.
10. And here donors and recipients are tied up in an intimate secret that they collectively share.
11. Translation: 'Egg donation is an intimate thing, they won't talk.'
12. I found this idea quite strange at the time since it was clear to me that Israelis do talk about 'intimate' things in quite public ways. Already upon my arrival, and increasingly throughout the duration of my research, I watched countless television programmes about issues of reproduction, sex and the

body. The resonance with Lauren Berlant's (1997) critique of the intimate public sphere in the USA is pertinent here. Where identity politics are ever utterable, issues of class inequalities at the time of her writing were unspeakable. The ever-changing nature of politics and 'public spheres' is palpable, however, as was witnessed in the USA in the year of 2011, the Oakland General Strike (a one-day general strike that took place on 2 November 2011 in Oakland, California), which was part of the wider global Peoples' Occupy Movements spread across the world and that were meant to empower citizens to recognize social, political and economic inequalities.

13. These were references to the advanced age of women who conceived after long periods of time spent 'trying'.
14. This is of course also the case in other national contexts such as the USA (Becker, 2000), Britain (Franklin, 1997) and Egypt (Inhorn, 2003).
15. The exact details of this can be accessed online (in Hebrew) on the Israeli Ministry of Health's website, under the Committees section. This protocol was part of a series of meetings held in the year 2000 by the Israel Ministry of Health's Public–Professional Committee on the Matter of Egg Donation. It may be the case that some of these deliberations have informed the 2010 Egg Donation Law. It is Protocol 11a Public–Professional Committee on the Matter of Egg Donation (2000).
16. See Stop the Wall (2011) for information about the Security Fence.
17. There is a growing body of work that examines Muslim laws on reproductive technologies (see Clarke, 2009; Inhorn, 2003; Tremayne, 2006).
18. Like Maryam, many of the women and men whom I interviewed challenged the idea of 'free' IVF and widely available treatment.
19. Israelis also use this word, *'primitivi'im'*. The term was used early in Israeli history to connote 'Oriental' Jews who were perceived as backward. Interestingly this term has been adopted by some Mizrahim as a way of distancing themselves from 'the primitives' by taking on a hegemonic kind of border policing. This has also been noted by Rhoda Kanaaneh when she discusses Palestinians, population discourse in Israel and modernity (Kanaaneh, 2002).
20. Her parents came to Israel from India.
21. In order to avoid the possibility of 'adultery' (see Kahn, 2000).
22. Here, the specificity of Hebrew is important. She used the common term for 'in Israel', which is *ba-Aretz*. This literally means 'in the [L]and [of Israel]' and to me (now situating myself strategically as an outsider), this has biblical and Zionist connotations of 'the Greater Land of Israel' (this is the phrase that refers to the imagined borders of Israel reaching all the way to contemporary Iraqi borders). Yet on a daily basis this term is used to refer to the country, and those connotations go unnoticed.
23. This word is similar to the English notion of 'bastard'. However, *mamzer* has specific Jewish legal status and many social restrictions relating to marriage and burial. See Kahn (2000) for an in-depth discussion.
24. 'Commandments' or observing Jewish law.
25. This is the clinic that was allegedly involved in the famous egg-stealing case discussed in Chapter 6.
26. The four nationalized health insurance companies.
27. I did not directly approach these women myself because at the start of my research at this clinic, a protocol was arranged between myself and the

staff that I would not approach patients directly in order to maintain a discreet and non-invasive atmosphere. Thus the manager, nurses and staff generally facilitated interviews for me by approaching women and men in the clinic and by asking them if they would be willing to let me speak with them.

28. There is an extremely low percentage of Mizrahi women in academic posts in Israeli academia (Lavie, 2011b).
29. And see Meira Weiss's moving account of *Conditional Love* (1994), where Jewish Israeli parents were found to abandon children born with different kinds of 'deformities' or disabilities.
30. Being non-white is very ordinary in Israel, in fact a majority of Israelis are not of European Ashkenazi descent.
31. While rejecting the idea that she could accept an egg from a Palestinian or Arab woman, Rina self-identified as Moroccan and used Arabic words in daily life. *Ya'ani*, Arabic for, 'I mean' or 'like' in an Anglo-American manner of speaking, is a part of common everyday Israeli Hebrew for Ashkenazim and Mizrahim, as are many other Arabic words. For Rina and Albert and many other Israeli couples, their sense of selfhood and kinship or physical reproduction did not allow for receiving an egg from an 'Arab' (that which is thought to contain 'the Arab'). Yet everyday Israeli Hebrew reproduces, in many ways, the intimate relationship with the Arabic-speaking Palestinians with whom they coinhabit geographical space.
32. This debate clearly also indexes the tensions between anthropology as a faithful description of what people say and intervention or critique.
33. Critics of Israeli racial practices are minor voices within the larger Israeli society and they face resistance even from the, mostly Ashkenazi, Israeli left (Lavie, 1996).
34. Geography intersects with race and class because new Jewish migrants from other parts of the Middle East and North Africa were traditionally settled in regions bordering Palestinian territories.
35. The Law of Return, first enacted in 1950, encodes the right of every Jew to 'return' to Israel, whereupon they may be granted citizenship. Determining who is a Jew is unclear in the 1950 version of the law (Dominguez, 1989: 171).
36. According to the song, this mixture will lead, ultimately, to peace.
37. In North America this encouragement takes the form of free-of-charge (paid for by Zionist charitable organizations such as the United Jewish Appeal), organized trips to Israel that are both educational and recreational. It also takes the form of short courses on the Israeli military that have both practical (shooting practice, living in barracks, wearing uniform) and educational (classes on Zionism, Israel's wars, geography) aspects.
38. While privileging information about the identity of one's parents and a brief life history (including synagogues attended and community links) is often sufficient, more and more care is taken by Orthodox rabbis to ensure the 'purity' of the Jewish line. This purity is proven by reference to lineage and relations, rather than by any sort of blood or genetic test. Yet now the new Egg Donation Law dictates no cross-religion egg donation, there has been a hardening of the relation between religion and biology if not necessarily 'genetics'.

5 Borders: IVF @ the Vanguard of Technology + War

1. Behar (2008) writes about the ethnographer as a vulnerable observer.
2. Here I am invoking Donna Haraway's (1991) call to feminists to create feminist cyborg fictions as part of a politics of 'truth telling' about science and technology and their enmeshment in the production of various kinds of inequalities and subordinations.
3. This is a pseudonym. I have chosen the name Barukh, which means 'blessed' to indicate my respect for this supportive and encouraging teacher. Whatever our differences were during the research, I was grateful for his guidance and education.
4. See Chapter 4 for an analysis of 'mixture' and 'difference'.
5. Saba Mahmood (2005) and Talal Asad (2003) have written on the problematic nature in which academics have tackled secularism in the Middle East, as involving a particular kind of relationship between the 'individual' and the 'universal'. They have critiqued the representation of secularism and religious fundamentalism in Islamic countries. In Israel, Judaism has been represented as a 'civil religion' (Liebman and Don-Yehiya, 1983) and IVF technologies, policies and practices have often been attributed a particularly 'Jewish' character. However, despite the intimate links between the Jewish religion and the state in Israel, the default option of designating things and people in Israel as religious or secular or an interesting 'mix' of the two misses the opportunity to see when discussions of secularism and religion are also doing 'nationalism work' (Ben Porat, 2000).
6. One of the most persistent images of Palestinians in Israel during this time was as 'suicide bombers'. Thus, if a newspaper article opened with a comment that a 'Palestinian man entered [...]', it was usually followed by horrific descriptions of explosions, bodies and death. Inserting this statement here is a deliberate counter-narrative. Palestinians were to be found in IVF clinics as patients, doctors (at least one well-known Palestinian citizen of Israel is also an obstetrician-gynaecologist), nurses, scientists and cleaners.
7. This description of Barukh, listing his 'vital statistics', family, ethnicity, religiosity and geographic locale identify him as an individual of complex affiliations and identifications. As a former army colonel and as a professor and father he has a high rank in Israeli society. Yet he is also a Mizrahi man and lives in disputed territory since his arrival in Israel as a child. This aspect of his identity also intersects with the state's expansion in the Jerusalem region, settling Mizrahim in the more violent and contentious zones. Mizrahim were and still are often the front lines of Israeli expansion. Clearly Barukh, a high-powered, male doctor, cannot be seen simplistically as a victim of the state in this case. Nevertheless, the intersectionality of race, class and gender is important to bear in mind when thinking about Israeli borders.
8. There is a large body of literature on the insider/outsider status of the researcher, for example Narayan (1993).
9. The profoundly heteronormative status of Israeli IVF was obvious from the start. Although it was common to see women alone at an IVF clinic (on single women and IVF, see Kahn, 2000), I did not encounter any women who openly identified as lesbian, bisexual or queer. If there were any, perhaps the

strongly heteronormative culture prevented them from revealing this to me. Of course, Israel is not the only place where we find such heteronormativity. Also, such heteronormativity shifts over time, where according to some the Israeli government is now using its pro-gay stance to 'pinkwash' the occupation (Puar, 2011).

10. The ease of patient access to their medical files was surprising to me. Coming from Canada, where a person has to overcome immense hurdles to access their own medical files, this is a particularity that I noticed almost immediately. This ease of access was apparent throughout the research in different Israeli clinics that I observed.
11. As this is a teaching hospital, patients would expect there to be students observing and learning during their appointments. I was granted permission for such observation by the head of the unit and by a Ministry of Health research ethics board. Below I discuss one aspect of getting ethics approval, which is about the translation of my research aims to Professor Barukh.
12. Informed consent was obtained from the hospital itself through an application to the hospital's ethics board. Since this is a teaching hospital, patients sign one consent form for all of the studies and observations of students. My project was subsumed under the consent received in that form.
13. The references are many, for two examples see Sperling (2010) and Kahn (2000).
14. The IDF's actions in Palestinian towns is almost always portrayed as a reaction to violence incurred, rather than planned aggression that results in Palestinian 'victims' responding.
15. It is interesting that he felt that the details were relevant. He did not seem to be merely gossiping but rather was 'helping' by giving me a 'social' account.
16. See Bharadwaj (2009) for another perspective on 'disposal' in which citizen/subjects are sacrificed in the service of a booming economy.
17. For a similar example of the fear of Muslim over-reproduction in India, see Jeffery and Jeffery (2006).
18. For a thorough and engaging discussion of the links between demography and Israel, see Portugese (1998).
19. 'Legitimacy' is an important aspect of Israeli immigration. Not only is the 'quantity' of immigrants important, but the 'quality' of those immigrants is equally crucial (as mentioned in Chapter 1 and as I discuss further in Chapter 4).
20. I am aware that there was only ever a partial acceptance of my presence. Many continued to regard me with caution.
21. This type of response is common to biomedical personnel in other contexts such as those in the UK.
22. I use this 'identity' cautiously. As can be seen from my fieldnote extracts throughout this book, this was a position very much in process. In Israel, by revealing any anti-occupation (of Palestine) views one automatically brands oneself as a 'leftist'. While sitting at a *shiva* (Jewish mourning ritual) for my great-aunt who died, I heard one distant relative remark that 'those leftists' are just like 'the Kapos' (Jews who aided Nazis in return for the promise of a reduced punishment). 'They're bad blood,' he said.

23. Today there are civil laws against boycotting Israeli goods or voicing vociferous criticism of Israeli organizations.
24. See Chapters 1 and 7 for a discussion of the ethics of such critical representations of Israelis. Whilst I am aware that Barukh might be angry about being represented in this manner, I am also conscious that a political critique of the daily operation of the ways in which Zionism and egg donation make one another palpable is facilitated by these detailed discussions.
25. Occasionally when Barukh was elsewhere and there was a shortage of nurses, the medical resident would conduct a *dikkur* and ask me to help him prepare the surgical table, maintaining its sterility. I was taught how to open packages of tubes and needles without touching them. I was also occasionally enlisted to pass on flasks containing extracted fluids to the laboratory.
26. These identifiers are noteworthy. Work, gender and ethnicity align in some stark ways, in a hierarchy with the doctor at the top, embryology and anaesthesia next, nursing and then cleaners at the bottom. Yet each 'layer' has different and shifting power. Nearly all of the physicians were men. There was one woman physician at the clinic, who was Ashkenazi. Although there were several Mizrahi male doctors, there were no women Mizrahi doctors. All of the embryologists, anaesthetists and nurses were women. This alignment of gender with status-linked jobs meant that men were in the more 'prestigious' roles and women in what were seen as 'assisting' or subsidiary roles. All of the embryologists were Ashkenazi Israeli-born women, while anaesthetists were Eastern European and Russian women and all of the nurses were Mizrahi women and Palestinian men. Furthermore, all of the cleaners of the surgical ward were Palestinian and Mizrahi women and men. While embryology is a respected science in Israel, one embryologist whom I interviewed claimed that it was not respected as its own Profession and in his opinion it thus tended to be more of a 'feminized' job that women (and at other clinics, foreign men) did. Its association with a lower status made embryology populated by fewer Ashkenazi men. Anaesthesia posts seemed to be held by people who had obtained medical degrees in other countries, where such degrees were not recognized in Israel. These were often women who self-identified as Russian. These striking divisions of labour along gender and racial lines indicate the connections between 'race' and class in Israeli society that are an overriding concern for many Mizrahi feminists (Azot Metsach, 2002; and Achoti).
27. During one oocyte extraction, the anaesthetist asked Barukh if they turn down the lights in order to make it seem 'more like the real thing', that is, sexual intercourse.
28. 'Seeing' these holes is difficult and requires 'skill'. After much practice, I was eventually able to make out black circles that I then interpreted as follicles. But this is not straightforward since blood vessels also appear as black circles on the monitor. I asked Barukh how he distinguishes these. He replied that one learns to discern the difference with practice. An experienced and senior professor was commissioned by one woman to conduct her oocyte aspiration and he had such difficulty in finding the follicles, that in the midst of the procedure he handed over the probe to Barukh who completed the procedure for him.
29. This is the word for an attack, which literally means 'an injury'.

30. Rehavam Ze'evi, then leader of the ultra-nationalist Moledet Party had been advocating 'transfer' since the late 1970s, and in July 2001 echoed his calls for the ethnic cleansing of Palestinians from Israel, the West Bank and Gaza (Edelman, 2007).
31. A Jewish settlement in the West Bank.
32. 'Martyr' in Arabic. It is also another name for suicide bombers.
33. The feminized and infantilized space of the laboratory, decorated with cartoon stickers and filled with conversations of mothers with their children, women with their partners and discussions of summer camp, was equally a place where some of the most highly technical and complex work on IVF was done.
34. My thanks to Donna Haraway for posing the question to me of what is the gender of the egg (Donna Haraway, personal communication, 2002).
35. The embryologist explained that the zona was the transparent outer layer of membrane around a mature oocyte.
36. I attempted to contact Haim and his wife in the coming weeks but my calls were not answered. Neither Haim nor his wife ever contacted me.
37. I would like to thank Yehudit Keshet, with whom I had many discussions on the topic and whose book *Checkpoint Watch* (2006) was very informative on this matter.
38. And see Eyal Weizman's (2007) stunning analysis of the architectural and geographic dimensions of the Israeli occupation of Palestine.

6 ExplosionCrisis: 'The Episode that Exploded', Women's Bodies and Nationalism

1. At the time of this research only women involved in IVF treatments could donate eggs. In May 2010, a law was passed permitting the payment of donors not themselves involved in IVF treatments.
2. It is unclear what her particular religious reasoning was, as most religious Jewish authorities in Israel permit ova donation.
3. In private clinics, IVF and related services can be paid for out of pocket. Most *kupot holim* do not cover all of the costs of ova donation. For more on the privatization of Israeli healthcare see Filc (2009).
4. Interview with one of the investigating team of the parliamentary group assigned to the case, Jerusalem, 2002.
5. The Hebrew word for this would be *mishar* (commerce). The usage of *mashara* is one example of the common usage of Arabic words or 'Easternized' versions of Hebrew words.
6. This phrase was used by Israeli Prime Minister Ariel Sharon on a televized address to the nation on 31 March 2002. This war, Sharon continued, was against the *tashtit ha-terror*, the 'terror infrastructure'. These metaphors of land and home are common tropes in nationalism (Anderson, 1991 [1983]).
7. Having attended a Zionist school in Canada, I learned at school of the many wars that put the Israeli state's very existence in 'crisis'. Yet, accompanying these narratives of crisis would be triumphal stories of the state's army winning wars against a stronger enemy.

8. Presaging this in his 'Theses on the philosophy of history' (1999 [1947]), Walter Benjamin comments that 'crisis' and emergency are persistent. Reflecting on Jews living in Europe under Fascism, he states, '[t]he tradition of the oppressed teaches us that the "state of emergency" in which we live is not the *exception* but the *rule*' (ibid.: 248, my italics). The context of Benjamin's comments is very specific. He refers to Jews living under Fascism. Yet, the ideas he expressed in 1940, when he originally wrote his 'Theses on the philosophy of history', resonates with Israeli self-representation as a Jewish nation in crisis, whether this be a demographic crises (Yuval-Davis, 1997; Portugese, 1998), or a crisis of war as indicated by the quotation above from former Israeli Prime Minister Ariel Sharon. It is important to note, however, that Benjamin was talking about 'the oppressed', and not about Jews in general. This is a vital distinction.
9. Groups such as Courage to Refuse, made up of Israeli soldiers from high to low ranks, were vilified and some of their members who refused to serve in the Occupied Territories were jailed (see Seruv, 2012).
10. Some of these media accounts can be found in the following Israeli newspaper articles by: Kraw (2000a, 2000b), Reznick (2000a, 2000b), Reznick and Kraw (2000), Reznick and Shadmi (2001a, 2001b), Rozenblum (2000a, 2000b) and Siegel-Itzkovitch (2000).
11. This would be the first explicit egg donation legislation in Israel. Egg donation was previously regulated through the 1987 IVF Law.
12. My account draws heavily on Balaban's account, which was narrated to me both in formal interviews and in casual conversations with her and her husband Itamar Kastelansky. Balaban was one of my primary research participants. I spent several hours interviewing her, chatting with her over coffee in her home and 'hanging out' at her house with her family. Balaban founded CHEN, the primary patient advocacy and support group for egg recipients in Israel. The name of her organization belonged to her, now deceased, father. As previously noted, the word *chen* means 'charm', and a leading Israeli anthropologist pointed out that the acronym 'CHEN' is used in the Israeli military to mean women's brigade. Balaban is well connected within the Israeli media, medical and political communities and hence made a very good, professional interlocutor.
13. A record of these meetings (in Hebrew) is found at Public–Professional Committee on the Matter of Egg Donation (2001).
14. This proposed law can be found (in Hebrew) at ibid.
15. Physicians lobbied against having what they saw as the intrusion by religious authorities on what they regarded as their own domain. Therefore after much political argument it was agreed that the Ministry of Justice would house the donor registry. This was seen as a fair and impartial location that would not give preference to any of the religious parties, secular or medical interests.
16. This was also part of Israel's attempt to 'be global', or to keep up with Western ideals of what people 'should' have available to them in the way of technology and medical treatments. Unfortunately I do not have the space here to develop this argument.
17. These are only part of the motivating factors of each committee. Each politician was also driven by many other factors, including their own career trajectories.

18. Ova cannot survive outside the human body for longer than a few hours. After this period they become unviable. Fertilized ova, however, can be cryopreserved and later thawed for use in IVF.
19. The High Court of Justice.
20. Mira Heubner Har-El, Legal Adviser, Israel Ministry of Health, personal communication, 24 June 2002.
21. Ben-Porat (2000: 223) further states, 'When operating within the framework of a national discourse, secularism often relies on religious foundations; it has, therefore, a far more complex relationship with religion than most secularists would be willing to admit – one that often undermines the ethos of equality and freedom.'
22. *Hul* is an acronym for the words *huts la-Aretz*, meaning 'outside the country', but the acronym is used as an everyday word.
23. Similar discourses were mobilized throughout the meeting 'protocols' (transcripts) of Halperin's committee (see Public–Professional Committee on the Matter of Egg Donation, 2001).
24. Monica Konrad (1998: 42) argues that an ovum is a woman's body exteriorized.
25. It may be relevant that the Palestinians whom I interviewed did not narrate egg donation as being in crisis. Since they are not considered 'legitimate' citizens of the state (if we take citizenship to be broadly defined as a cultural thing and not just about official state recognition), they might not be inclined to provide me with a national story (it may be that they provided me with another kind of national story, which would be interesting to explore further).
26. In a sense, the state's rhetoric relies on a conflation of anti-Zionism with anti-Semitism (Butler, 2004).

References

Abu-Lughod, L. (1993) *Writing Women's Worlds: Bedouin Stories* (Berkeley, CA: University of California Press).
Agamben, G. (1995) *Homo Sacer: Sovereign Power and Bare Life* (Stanford, CA: Stanford University Press).
Ahad Ha'Am (1889) 'This is not the Way (The Wrong Way) (Lo zu Haderech)', originally published in Hebrew in *Ha-Melitz*, Zionism and History, *Zionism on the Web*, English translation, available at: http://www.zionismontheweb.org/ehad_haam1.htm (accessed 3 September 2011).
Ahmed, S. (2000) *Strange Encounters: Embodied Others in Post-Coloniality* (London: Routledge).
Ahmed, S. (2004) 'Affective economies', *Social Text*, 22, 2: 117–23.
Alcalay, A. (1993) *After Jews and Arabs: Remaking Levantine Culture* (Minneapolis, MN: University of Minnesota Press).
Almog, O. (2000) *The Sabra: The Making of the New Jew* (Berkeley, CA: University of California Press).
Amir, D. and O. Benjamin (1997) 'Defining encounters: who are the women entitled to join the Israeli collective?', *Women's Studies International Forum, Special Issue on Israel*, 20, 5: 639–50.
Anderson, B. (1991 [1983]) *Imagined Communities* (London and New York: Verso).
Appadurai, A. (1991) 'Global ethnoscapes: notes and queries for a transnational anthropology', in R. Fox (ed.), *Recapturing Anthropology: Working in the Present* (Santa Fe, NM: School of American Research Press), pp. 191–210.
Appadurai, A. (1996) *Modernity at Large: Cultural Dimensions of Globalization* (Minneapolis, MN: University of Minnesota Press).
Appadurai, A. (2002) 'Disjuncture and difference in global cultural economy', in J. X. Inda and R. Rosaldo (eds), *The Anthropology of Globalization: A Reader* (Oxford: Blackwell Publishing), pp. 46–63.
Aretxaga, B. (2003) 'Maddening states', *Annual Review of Anthropology*, 32: 393–410.
Arora, A. and A. Gambardella (2005) *From Underdogs to Tigers: The Rise and Growth of the Software Industry in Brazil, China, India, Ireland, and Israel* (Oxford: Oxford University Press).
Asad, T. (2003) *Formations of the Secular: Christianity, Islam, Modernity* (Stanford, CA: Stanford University Press).
Auerbach, E. (1957) *Mimesis: The Representation of Reality in Western Literature* (Princeton, NJ: Princeton University Press).
Avraham, E. and A. First (2003) '"I buy American": the American image as reflected in Israeli advertising', *Journal of Communication*, 53, 2: 282–99.
Achoti [organization] Was available at: http://www.achoti.org.il (website has been suspended and is currently inaccessible).
Azot Metsach (ed.) (2002) *Mizrahi Feminism*, vol. 3 (Jerusalem: Justice Unit – Students for Social Justice) (Hebrew).

B'Tselem, Israeli Information Center for Human Rights in the Occupied Territories, Jerusalem, the general page for testimonies does not have a date of publication, it is constantly updated, available at: http://www.btselem.org

B'Tselem, Israeli Information Center for Human Rights in the Occupied Territories, Jerusalem (2002a) 'IDF causes death of infant by preventing mother's evacuation to hospital', Testimony by T. Fatuah, April, available at: http://www.btselem.org/testimonies/20020413_death_of_tahani_fatouhs_premature_baby (accessed 29 December 2011).

B'Tselem, Israeli Information Center for Human Rights in the Occupied Territories, Jerusalem (2002b) 'IDF rubber bullet causes head injury to seven-month pregnant woman', Testimony by Suheir Shhada, 8 July, available at: http://www.btselem.org/testimonies/20020708_injury_of_suheir_shhada (accessed 29 December 2011).

B'Tselem, Israeli Information Center for Human Rights in the Occupied Territories, Jerusalem (2002c) 'IDF shoots Palestinian woman in labor while on her way to hospital', February, Testimony by Shadiyeh Fahmi Khaled 'Odeh, available at: http://www.btselem.org/testimonies/20020224_huwwara_roadblock_witness_odeh (accessed 29 December 2011).

B'Tselem, Israeli Information Center for Human Rights in the Occupied Territories, Jerusalem (2002d) 'IDF shoots Palestinian woman in labor while on her way to hospital, February 2002', 25 February, Testimony, available at: http://www.btselem.org/testimonies/20020225_huwwara_roadblock_witness_al_hayek (accessed 29 December 2011).

B'Tselem, Israeli Information Center for Human Rights in the Occupied Territories, Jerusalem, [2011], available at: http://www.btselem.org/testimonies (accessed 29 December 2011).

Back, L. and J. Solomos (eds) (2000) *Theories of Race and Racism: A Reader* (London: Routledge).

Barad, K. (2007) *Meeting the Universe Halfway: Quantum Physics and the Entanglement of Matter and Meaning* (Durham, NC: Duke University Press).

Becker, G. (1997) *Disrupted Lives: How People Create Meaning in a Chaotic World* (Berkeley, CA: University of California Press).

Becker, G. (2000) *The Elusive Embryo: How Women and Men Approach New Reproductive Technologies* (Berkeley, CA: University of California Press).

Behar, R. (2008) *The Vulnerable Observer: Anthropology that Breaks Your Heart* (Boston, MA: Beacon Press).

Ben-Porat, G. (2000) 'A state of holiness: rethinking Israeli secularism', *Alternatives*, 25, 2: 223–45.

Benjamin, W. (1999 [1947]) 'Theses on the philosophy of history', in H. Arendt (ed.), *Illuminations: Walter Benjamin* (London: Pimlico), pp. 245–55.

Berkovitch, N. (1997) 'Motherhood as a national mission: the construction of womanhood in the legal discourse in Israel', *Women's Studies International Forum*, 20, 5–6: 605–19.

Berlant, L. (1997) *Queen of America Goes to Washington City: Essays on Sex and Citizenship* (Durham, NC: Duke University Press).

Bernstein, D. (1980) 'Immigrants and society: a critical view of the dominant school of Israeli society', *British Journal of Sociology*, 31, 2: 246–64.

Bharadwaj, A. (2003) 'Why adoption is not an option in India: the visibility of infertility, the secrecy of donor insemination, and other cultural complexities', *Social Science and Medicine*, 56, 9: 1867–80.

Bharadwaj, A. (2006a) 'Sacred modernity: religion, infertility, and technoscientific conception around the globe', *Culture, Medicine, Psychiatry*, 30, 4: 423–5.

Bharadwaj, A. (2006b) 'Sacred conceptions: clinical theodicies, uncertain science, and technologies of procreation in India', *Culture, Medicine, Psychiatry*, 30, 4: 451–65.

Bharadwaj, A. (2009) 'Assisted life: the neoliberal moral economy of embryonic stem cells in India', in D. Birenbaum-Carmeli and M. C. Inhorn (eds), *Assisting Reproduction, Testing Genes: Global Encounters with New Biotechnologies* (New York: Berghahn Books), pp. 239–57.

Bick, E. (1997) 'Ovum donations: a rabbinic conceptual model of maternity', in E. Feldman and J. B. Wolowelsky (eds), *Jewish Law and the New Reproductive Technologies* (Hoboken, NJ: Ktav Publishing House), pp. 83–105.

Billig, M. (1995) *Banal Nationalism* (London: Sage).

Birenbaum-Carmeli, D. (1998) 'Reproductive partners: doctor–woman relations in Canadian and Israeli IVF contexts', in N. Scheper-Hughes and C. Sargeant (eds), *Small Wars: The Cultural Politics of Childhood* (Berkeley, CA: University of California Press), pp. 75–92.

Birenbaum-Carmeli, D. and Y. Carmeli (2002) 'Physiognomy, familism and consumerism: preferences in donor insemination', *Social Science and Medicine*, 54, 3: 363–76.

Bleich, D. J. (1997) '*In vitro* fertilization: questions of maternal identity and conversion', in E. Feldman and J. B. Wolowelsky (eds), *Jewish Law and the New Reproductive Technologies* (Hoboken, NJ: Ktav Publishing House), pp. 106–14.

Blyth, E., (1998) 'Donor assisted conception and donor offspring rights to genetic origins information', *International Journal of Children's Rights*, 6, 3: 237–53.

Bordo, S. (1993) 'Reading the slender body', in *Unbearable Weight: Feminism, Western Culture and the Body* (Berkeley, CA: University of California Press), pp. 185–212.

Bourdieu, P. (1977) *Outline of a Theory of Practice* (Cambridge: Cambridge University Press).

Bowen, J. (2007) *Why the French Don't Like Headscarves: Islam, the State and Public Space* (Princeton, NJ: Princeton University Press).

Boyarin, J. (1996) *Palestine and Jewish History: Criticism at the Borders of Ethnography* (Minneapolis, MN, and London: University of Minneapolis Press).

Brah, A. (2003) 'Diaspora, borders and transnational identities', in R. Lewis and S. Mills (eds), *Feminist Post-Colonial Theory: A Reader* (Edinburgh: Edinburgh University Press), pp. 613–34.

Braidotti, R. (1994) *Nomadic Subjects: Embodiment and Sexual Difference in Contemporary Feminist Theory* (New York: Columbia University Press).

Breznitz, D. (2005) 'Collaborative public space in a national innovation system: a case study of the Israeli military's impact on the software industry', *Industry and Innovation*, 12, 1: 31–64.

Butler, J. (1993) *Bodies that Matter: On the Discursive Limits of 'Sex'* (New York: Routledge).

Butler, J. (2004) *Precarious Life: The Powers of Mourning and Violence* (London: Verso).

Carmeli, Y. S. and K. Applebaum (eds) (2004) *Consumption and Market Society in Israel* (New York: Berg).

Castañeda, C. (2002) *Figurations: Child, Bodies, Worlds* (Durham, NC: Duke University Press).

Chehata, H. (2012) 'Israel: Promised Land for Jews ... as long as they're not Black', *Race and Class*, 53, 4: 67–77.

Chernichovsky, D. (1992) 'Health system reforms in industrialized democracies: an emerging paradigm', *Millbank Quarterly*, 73, 3: 339–72.

Chernichovsky, D. and D. Chinitz (1995) 'The political economy of healthy system reform in Israel', *Health Economics*, 4, 2: 127–41.

Clarke, M. (2009) *Islam and New Kinship: Reproductive Technology and the Shariah in Lebanon* (London: Berghahn Books).

Clifford, J. (1986) 'Introduction: partial truths', in J. Clifford and G. E. Marcus (eds), *Writing Culture: The Poetics and Politics of Ethnography* (Berkeley, CA: University of California Press), pp. 1–26.

Clifford, J. and G. Marcus (1986) *Writing Culture: The Poetics and Politics of Ethnography* (Berkeley, CA: University of California Press).

Cohen, L. (2005) 'Bioavailability and exception', in A. Ong and S. J. Collier (eds), *Global Assemblages: Technology, Politics, and Ethics as Anthropological Problems* (Oxford: Blackwell Publishing), pp. 79–90.

Cohen, L. (2011) 'Migrant supplementarity: remaking biological relatedness in Chinese military and Indian five-star hospitals', *Body & Society*, 17, 2–3: 31–54.

Colen, S. (1995) '"Like a mother to them": stratified reproduction and West Indian childcare workers and employers in New York', in F. Ginsburg and R. Rapp (eds), *Conceiving the New World Order: The Global Politics of Reproduction* (Berkeley, CA: University of California Press), pp. 78–102.

Comaroff, J. and J. Comaroff (2000) 'Millenial capitalism: first thoughts on a second coming', *Public Culture*, 12, 2: 291–343.

Corrigan, O. (2003) 'Empty ethics: the problem with informed consent', *Sociology of Health and Illness*, 25, 7: 768–92.

Dickenson, D. (2006) 'The lady vanishes: what's missing from the stem cell debate', *Journal of Bioethical Inquiry*, 3, 1–2: 43–54.

Dickenson, D. (2007) *Property in the Body: Feminist Perspectives* (Cambridge: Cambridge University Press).

Dolgin, J. (1995) 'Family law and the facts of the family', in S. Yanagisako and C. Delaney (eds), *Naturalizing Power: Essays in Feminist Cultural Analysis* (New York and London: Routledge), 47–68.

Dominguez, V. (1989) *People as Subject, People as Object: Selfhood and Peoplehood in Contemporary Israel* (Madison, WI: University of Wisconsin Press).

Douglas, M. (1966) *Purity and Danger: An Analysis of Pollution and Taboo* (London: Routledge).

Downey, G. and J. Dumit (eds) (1995) *Cyborgs and Citadels: Anthropological Interventions in Emerging Sciences and Technologies* (Santa Fe, NM: School of American Research Press).

Duden, B. (1993) *Disembodying Women: Perspectives on Pregnancy and the Unborn* (Cambridge, MA: Harvard University Press).

Dyer, R. (2000) 'The matter of whiteness', in L. Back and J. Solomos (eds), *Theories of Race and Racism: A Reader* (London: Routledge), pp. 539–48.

Edelman, O. (2007) 'Rehavam Ze'evi assassin jailed for two life terms plus 100 years', *Haaretz* (3 December), available at: http://www.haaretz.com/news/

rehavam-ze-evi-assassin-jailed-for-two-life-terms-plus-100-years-1.234461 (accessed 19 September 2012).
El Haj, N. (2007) 'The genetic reinscription of race', *Annual Review of Anthropology*, 36: 283–300.
Eurostat (2007a) 'Population and social conditions: labour market', available at: http://epp.eurostat.ec.europa.eu/portal/page/portal/labour_market/introduction (accessed 13 August 2012).
Eurostat (2007b) 'Structural indicators: employment', available at: http://epp.eurostat.ec.europa.eu/portal/page/portal/structural_indicators/indicators/mployment (accessed 13 August 2012).
Even, D. (2011) 'Study: IVF treatment is doubling number of preemie twins', *Haaretz* (19 September), available at: http://www.haaretz.com/print-edition/news/study-ivf-treatment-is-doubling-number-of-preemie-twins-1.385242 (accessed 18 September 2012).
Feldman, E. and J. B. Wolowelsky (eds) (1997) *Jewish Law and the New Reproductive Technologies* (Hoboken, NJ: Ktav Publishing House).
Ferguson, J. (2002) 'Spatializing states: toward an ethnography of neoliberal governmentality', *American Ethnologist*, 29, 4: 981–1002.
Filc, D. (2005) 'The health business under neo-liberalism: the Israeli case', *Critical Social Policy*, 25, 2: 180–97.
Filc, D. (2009) *Circles of Exclusion: The Politics of Healthcare in Israel* (Ithaca, NY: Cornell University Press).
Fishman, R. H. B. (1998) 'Reproductive rights and risks in Israel', *The Lancet*, 352, 9158: 1998.
Fishman, R. H. B. (1999a) '*In-vitro* fertilisation in Israel', *The Lancet*, 353, 9170: 2163.
Fishman, R. H. B. (1999b) 'Infertility doctors use egg donors worldwide', *The Lancet*, 353, 9154: 736.
Fishman, R. H. B. (1999c) 'Israeli health minister advises access to unlimited *in-vitro* fertilisation', *The Lancet*, 353, 9152: 570.
Fishman, R. H. B. (2000) 'Israel faces changes in IVF regulations', *The Lancet*, 355, 9219: 1979.
Fortier, A.-M. (2001) 'New intimacies and un/marked (hetero)sexuality: intersections of ethnicity and sexuality in new multiculturalist Britain', Lancaster University, available at: http://www.comp.lancs.ac.uk/sociology/research/resalph.htm (accessed 5 January 2005).
Fortier, A.-M. (2005) 'Pride politics and multiculturalist citizenship', *Ethnic and Racial Studies*, 28, 3: 559–78.
Foucault, M. (1978) *The History of Sexuality, Volume 1: An Introduction* (New York: Vintage Books).
Frankenburg, R. (2000) 'White women, race matters: the social construction of whiteness', in L. Back and J. Solomos (eds), *Theories of Race and Racism: A Reader* (London: Routledge), pp. 447–61.
Franklin, S. (1997) *Embodied Progress: A Cultural Account of Assisted Conception* (London: Routledge).
Franklin, S. (2005) 'Stem Cells R Us: emergent life forms and the global biological', in A. Ong and S. J. Collier (eds), *Global Assemblages: Technology, Politics, and Ethics as Anthropological Problems* (Oxford: Blackwell Publishing), pp. 59–78.
Franklin, S., C. Lury and J. Stacey (2000) *Global Nature, Global Culture* (London: Sage Publications).

Freilin, R. (2000) 'Danger: AIDS in IVF outside Israel', *Ma'ariv* (17 December 2000) (Hebrew).

Freud, S. (1985 [1919]) 'The uncanny', Penguin Freud Library, *Volume 14: Art and Literature,* ed. by A. Dickson (London: Penguin Freud Library), pp. 335–76.

Gal, S. and G. Kligman (eds) (2000) *Reproducing Gender: Politics, Publics and Everyday Life after Socialism* (Princeton, NJ: Princeton University Press).

Gampel, Y. (1996) 'The interminable uncanny', in L. Rangell and R. Moses-Hrushovski (eds), *Psychoanalysis at the Political Border* (Madison, CT: International Universities Press), pp. 85–96.

Geertz, C. (1973) 'Thick description: toward an interpretive theory of culture', in C. Geertz (ed.), *The Interpretation of Culture* (New York: Basic Books), pp. 3–30.

Geertz, C. (1988) *Works and Lives: The Anthropologist as Author* (Cambridge: Polity Press).

Gibbon, S. and C. Novas (eds) (2007) *Genetics and the Social Sciences: Making Biosociality* (London: Routledge).

Gilman, S. (1991) *The Jew's Body* (New York: Routledge).

Ginsburg, F. and R. Rapp (1991) 'The politics of reproduction', *Annual Review of Anthropology,* 20: 311–43.

Ginsburg, F. and R. Rapp (eds) (1995) *Conceiving the New World Order: The Global Politics of Reproduction* (Columbia, CA: University of California Press).

Glick-Schiller, N., L. Basch and C. Blanc Szanton (1995) 'From immigrant to transmigrant: theorizing transnational migration', *Anthropology Quarterly,* 68, 1: 48–63.

Glick-Schiller, N., et al. (1992) 'Transnationalism: a new analytic framework for understanding migration', *Annals of the New York Academy of Sciences,* 645, 1: 1–24.

Goldberg, D. T. (2002) *The Racial State* (Oxford: Blackwell).

Goldberg, M. (2001) 'Approved: import of ova from outside Israel', *Yediot Aharonot* (25 December) (Hebrew).

Gordon, N. (2011) 'Israel's emergence as homeland security capital', in E. Zureik, D. Lyon and Y. Abu-Laban (eds), *Surveillance and Control in Israel/Palestine: Population, Territory and Power* (New York: Routledge), pp. 153–70.

Gross, N. (2001) 'Infertile women to benefit from a revolutionary Israeli bill', *The Jerusalem Report Magazine,* available at: http://www.jrep.com/Reporter/Article-15.html (accessed 16 November 2001).

Gupta, J. A. (2006) 'Towards transnational feminisms: some reflections and concerns in relation to the globalization of reproductive technologies', *European Journal of Women's Studies,* 13, 1: 23–38.

Gürtin, Z. (2011) 'Banning reproductive travel: Turkey's ART legislation and third-party assisted reproduction', *Reproductive Biomedicine Online,* 23, 5: 555–64.

Haimes, E. (1990) 'Recreating the family? Policy considerations relating to the "new" reproductive technologies', in M. McNeil, I. Varcoe and S. Yearly (eds), *The New Reproductive Technologies* (Basingstoke: MacMillan), pp. 154–72.

Haimes, E. (1993) 'Theory and methodology in the analysis of the policy process: a case study of the Warnock Committee on Human Fertilization and Embryology', in M. Hill (ed.), *New Agendas in the Study of the Policy Process* (Hemel Hempstead: Harvester Wheatsheaf), pp. 152–75.

Hall, S. (1991) 'The local and the global: globalization and ethnicity', in A. D. King (ed.), *Culture, Globalization and the World-System* (London: Macmillan), pp. 41–68.

Handelman, D. (2004) *Nationalism and the Israeli State: Bureaucratic Logic in Public Events* (Oxford: Berg).
Handler, R. and D. Segal (1990) *Jane Austen and the Fiction of Culture* (Tucson, AZ: University of Arizona Press).
Haraway, D. (1991) *Simians, Cyborgs and Women: The Reinvention of Nature* (New York: Routledge).
Haraway, D. (1997) *Modest_Witness@Second_Millenium. FemaleMan©_Meets_OncomouseTM, Feminism and Technoscience* (New York: Routledge).
Haraway, D. (2003) *The Companion Species Manifesto: Dogs, People, and Significant Otherness* (Chicago, IL: Prickly Paradigm Press).
Haraway, D. (2004) *The Haraway Reader* (New York: Routledge).
Hayden, C. P. (1995) 'Gender, genetics and generation: reformulating biology in lesbian kinship', *Cultural Anthropology*, 10, 1: 41–63.
Hord, C., et al. (1991) 'Reproductive health in Romania: reversing the Ceausescu legacy', *Studies in Family Planning*, 22, 4: 231–40.
Howell, J. (1996) *Technology in the Hospital: Transforming Patient Care in the Early Twentieth Century* (Baltimore, MD: Johns Hopkins University Press).
Hubert, H. and M. Mauss (1964 [1899]) *Sacrifice: Its Nature and Function* (Chicago, IL: University of Chicago Press).
Inhorn, M. (2003) *Local Babies, Global Science: Gender, Religion, and* in Vitro *Fertilization in Egypt* (New York: Routledge).
Inhorn, M. (2011) 'Diasporic dreaming: return reproductive tourism to the Middle East', *Reproductive Biomedicine Online*, 23, 5: 582–91.
Inhorn, M. C. and P. Patrizio (2009) 'Rethinking reproductive "tourism" as reproductive "exile"', *Fertility and Sterility*, 92, 3: 904–6.
Ivry, T. (1999) 'Reproduction as martial art', paper presented at 34th World Congress of the International Institute of Sociology, Tel Aviv, 12–15 July.
Ivry, T. (2009) *Embodying Culture: Pregnancy in Japan and Israel* (New Brunswick, NJ: Rutgers University Press).
Jean-Klein, I. (2001) 'Nationalism and resistance: the two faces of everyday activism in Palestine during the Intifada', *Cultural Anthropology*, 16, 1: 83–126.
Jeffery, P. and R. Jeffery (2006) *Confronting Saffron Demography: Religion, Fertility and Women's Status in India* (New Delhi: Three Essays Collective).
Kahn, S. M. (2000) *Reproducing Jews: A Cultural Account of Assisted Conception in Israel* (Durham, NC: Duke University Press).
Kanaaneh, R. (2002) *Birthing the Nation: Strategies of Palestinian Women in Israel* (Berkeley, CA: University of California Press).
Keshet, Y. K. (2006) *Checkpoint Watch: Testimonies from Occupied Palestine* (London: Zed Books).
Khanna, R. (2007) 'Indignity', *Ethnic and Racial Studies*, 30, 2: 257–80.
Kligman, G. (1992) 'Abortion and international adoption in Ceausescu's Romania', *Feminist Studies*, 18, 2: 405–19.
Kligman, G. (1998) *The Politics of Duplicity: Controlling Reproduction in Ceausescu's Romania* (Berkeley, CA: University of California Press).
Konrad, M. (1998) 'Ova donation and symbols of substance: some variations on the theme of sex, gender and the partible person', *Journal of the Royal Anthropological Institute*, 4: 643–67.
Konrad, M. (2005) *Narrating the New Predictive Genetics: Ethics, Ethnography and Science* (London: Cambridge University Press).

Kraw, N. (2000a) 'The police: the gynecologists who were arrested extracted eggs in quantities for trade', *Haaretz* (17 May), available at: http://www.haaretz.co.il (accessed 23 January 2001) (Hebrew).

Kraw, N. (2000b) 'Two gynecologists are suspected of endagering the lives of women patients', *Haaretz* (17 May), available at: http://www.haaretz.co.il (accessed 23 January 2001) (Hebrew).

Kuntsman, A. (2009a) *Figurations of Violence and Belonging: Queerness, Migranthood and Nationalism in Cyberspace and Beyond* (Bern: Peter Lang).

Kuntsman, A. (2009b) 'The currency of victimhood in uncanny homes: queer immigrants' claims for home and belonging through anti-homophobic organising', *Journal of Ethnic and Migration Studies*, 35, 1: 133–49.

Lambek, M. (2011) 'Catching the local', *Anthropological Theory*, 11, 2: 197–221.

Landau R. (1998) 'The management of genetic origins: secrecy and openness in donor assisted conception in Israel and elsewhere', *Human Reproduction*, 13, 11: 3268–73.

Lavie, S. (1996) 'Blowups in the borderzones: Third World Israeli authors' gropings for home', in S. Lavie and T. Swedenburg (eds), *Displacement, Diaspora, and Geographies of Identity* (Durham, NC: Duke University Press), pp. 55–96.

Lavie, S. (2011a) 'Staying put: crossing the Israel–Palestine border with Gloria Anzaldúa', *Anthropology and Humanism*, 36, 1:101–21.

Lavie, S. (2011b) 'Mizrahi feminism and the question of Palestine', *Journal of Middle East Women's Studies*, 7, 2: 56–88.

Lentin, A. and R. Lentin (eds) (2008) *Race and State* (Newcastle: Cambridge Scholars Publishing).

Liebes, T. (2003) *American Dreams, Hebrew Subtitles: Globalization from the Receiving End* (Cresskill, NJ: Hampton Press).

Liebman, C. and E. Don-Yehiya (1983) *Civil Religion in Israel: Traditional Judaism and Political Culture in the Jewish State* (Berkeley and Los Angeles, CA: University of California Press).

Linder-Ganz, R. and N. Darom (2008) 'Step-children of the system', *Haaretz* (29 July), available at: http://www.haaretz.co.il/hasite/spages/1004785.html (accessed 30 January 2012) (Hebrew).

Lock, M. (1995) *Encounters with Aging Mythologies of Menopause in Japan and North America* (Berkeley, CA: University of California Press).

Lury, C. (2000) 'The united colours of diversity: essential and inessential culture', in S. Franklin, C. Lury and J. Stacey, *Global Nature, Global Culture* (London: Sage Publications), pp. 146–87.

Mahmood, S. (2005) *Politics of Piety: The Islamic Revival and the Feminist Subject* (Princeton, NJ: Princeton University Press).

Martin, E. (1991) *The Woman in the Body: A Cultural Analysis of Reproduction* (Boston, MA: Beacon Press).

Martin, E. (1995) 'From reproduction to HIV: blurring categories shifting positions', in F. Ginsburg and R. Rapp (eds), *Conceiving the New World Order: The Global Politics of Reproduction* (Columbia, CA: University of California Press).

Matorras, R. (2005) 'Reproductive exile versus reproductive tourism', *Human Reproduction*, 20, 12: 3571.

Mauss, M. (1990 [1950]) *The Gift: The Form and Reason for Exchange in Archaic Societies* (London: Routledge).

McNeil, M. (1993a) 'Editorial: procreation stories', *Science as Culture*, 3, 4: 477–82.

McNeil, M. (1993b) 'New reproductive technologies: dreams and broken promises', *Science as Culture*, 3, 4: 483–506.

McNeil, M., I. Varcoe and S. Yearley (1990) *The New Reproductive Technologies* (New York: St. Martin's Press).

Milner, E. (2001) 'The Medical Committee has decided: women volunteers, not undergoing IVF treatments, will be able to donate ova', *Yediot Aharonot* (20 March) (Hebrew).

Ministry of Foreign Affairs, Israel (2010) 'People: Minority Communities', 28 November, available at: http://www.mfa.gov.il/MFA/Facts+About+Israel/People/SOCIETY-%20Minority%20Communities (accessed 24 June 2011).

Mohanty, C. (1984) 'Under Western eyes: feminist scholarship and colonial discourses', *Boundary 2*, 12: 333–58.

Mol, A. (2002) *The Body Multiple: Ontology in Medical Practice* (London: Duke University Press).

Morgan, L. (2009) *Icons of Life: A Cultural History of Human Embryos* (Berkeley, CA: University of California Press).

Motzafi-Haller, P. (2004) 'Negotiating difference in Israeli scholarship: towards a new feminist discourse', in A. Kemp et al. (eds), *Israelis in Conflict: Hegemonies, Identities and Challenges* (Brighton: Sussex Academic Press), pp. 162–87.

Nahman, M. (2000) 'Embodied stories, pragmatic lives: intersex narratives on the Net' (unpublished MA thesis, York University, Toronto).

Nahman, M. (2005) 'Israeli extraction: an ethnographic study of ova donation and national imaginaries' (PhD dissertation, Lancaster University).

Nahman, M. (2006) 'Materialising Israeliness: difference and mixture in transnational ova donation', *Science as Culture*, 15, 3: 199–213.

Nahman, M. (2007) 'Synecdochic ricochets: biosocialities in a Jerusalem IVF clinic', in S. Gibbon and C. Novas (eds), *Biosocialities, Genetics and the Social Sciences: Making Biologies and Identities* (London: Routledge), 117–35.

Nahman, M. (2008) 'Nodes of desire: transnational egg sellers as theorists of reproduction in a neo-liberal age', *European Journal of Women's Studies*, 15, 2: 65–82.

Nahman, M. (2010) '"Embryos are our baby": abridging hope, body and nation in transnational ova donation', in J. Edwards, P. Harvey and P. Wade (eds), *Technologized Images, Technologized Bodies: Anthropological Approaches to a New Politics of Vision* (Oxford and New York: Berghahn Books), pp. 185–210.

Nahman, M. (2011) 'Reverse traffic: intersecting inequalities in human egg donation', *Reproductive Biomedicine Online*, 23, 5: 626–33.

Nahman, M. (2012) 'Making interferences: the cultural politics of transnational ova "donation"', in M. Knecht, M. Klotz and S. Beck (eds), *Reproductive Technologies as Global Form: Ethnographies of Knowledge, Practices and Transnational Encounters* (Frankfurt: Campus Verlag), pp. 305–30.

Narayan, K. (1993) 'How native is a "native" anthropologist?', *American Anthropologist*, 95, 3: 671–86.

Nitzan, J. and S. Bichler (2002) *The Global Political Economy of Israel* (London: Pluto Press).

Novas, C. and N. Rose (2000) 'Genetic risk and the birth of the somatic individual', *Economy and Society*, 29, 4: 485–513.

Novas, C. and N. Rose (2005) 'Biological citizenship', in A. Ong and S. J. Collier (eds), *Global Assemblages: Technology, Politics and Ethics as Anthropological Problems* (London: Blackwell), pp. 439–63.

Ochs, J. (2006) 'The politics of victimhood and its internal exegetes: terror victims in Israel', *History and Anthropology*, 17, 4: 355–68.

Ong, A. (1988) 'Colonialism and modernity: feminist re-presentations of women in non-Western societies', *Inscriptions*, 3, 4: 79–93.

Ong, A. (1999) *Flexible Citizenship: The Cultural Logics of Transnationality* (Durham, NC: Duke University Press).

Ong, A. and S. J. Collier (2005) *Global Assemblages: Technology, Politics and Ethics as Anthropological Problems* (London: Blackwell).

Paraszczuk, J. (2011) 'Four-year-old Filipina girl and her mother to be deported', *Jerusalem Post* (25 August), available at: http://jpost.com (accessed 30 January 2012).

Paxton, R. (ed.) (2004) *The Anatomy of Fascism* (London: Penguin).

Peled, Y. (2004) 'Profits or glory', *New Left Review*, 29: 47–72.

Pennings, G. (2002) 'Reproductive tourism as moral pluralism in motion', *Journal of Medical Ethics*, 28, 6: 337–41.

Pennings, G. (2004) 'Legal harmonization and reproductive tourism in Europe', *Human Reproduction*, 19: 2689–94.

Portugese, J. (1998) *Fertility Policy in Israel: The Politics of Religion, Gender, and Nation* (London: Praeger).

Prainsack, B. and Y. Hashiloni-Dolev (2009) 'Religion and nationhood: collective identities and the New Genetics', in P. Atkinson, P. Glasner and M. Lock (eds), *Handbook of Genetics and Society: Mapping the New Genomic Era* (London: Routledge), 404–21.

Project on Disney, The (1995) *Inside the Mouse: Work and Play at Disney World* (London: Rivers Oram Press).

Puar, J. (2011) 'Citation and censorship: the politics of talking about the sexual politics of Israel', *Feminist Legal Studies*, 19, 2: 133–42.

Public–Professional Committee on the Matter of Egg Donation (2000) 'Presentation of the Security Principles of the Data Protection System for Donated Eggs, Protocol 11a' (Israel: Ministry of Health Online), available at: http://www.old.health.gov.il/pages/default.asp?pageid=5105&parentid=5102&catid=6&maincat=1 (accessed 30 January 2012).

Public–Professional Committee on the Matter of Egg Donation (2001) 'Protocols of the Public–Professional Committee on the Matter of Egg Donation', Israeli Ministry of Health, available at: http://www.old.health.gov.il/pages/default.asp?maincat=1&catid=6&pageid=5102 (accessed 19 September 2012).

Rabinerson, D., et al. (2002) 'Subsidised oocyte donation in Israel (1998–2000): results, costs and lessons', *Human Reproduction*, 17: 1404–6.

Rabinow, P. (1995) 'Artificiality and enlightenment: from sociobiology to biosociality', in *Essays in the Anthropology of Reason* (Princeton, NJ: Princeton University Press), pp. 91–111.

Rabinow, P. (1999) *French DNA: Trouble in Purgatory* (Chicago, IL: University of Chicago Press).

Rapp, R. (1999) *Testing Women, Testing the Fetus: The Social Impact of Amniocentesis in America* (New York: Routledge).

Ravid, B. and D. Weiler-Polak, (2010) 'Cabinet approves deportation of 400 migrant children from Israel', *Haaretz* (8 January), available at: http://www.haaretz.com (accessed 30 January 2012).

Revel, M. (2002) 'The use of embryonic stem cells for therapeutic research', *Report of the Bioethics Advisory Committee*, Israel Academy of Sciences and Humanities, available at: http://www.academy.ac.il/data/reports_data/31/21.pdf (accessed 16 November 2012).

Reznick, R. (2000a) 'The lives of women patients were put at risk', *Haaretz* (17 May) (Hebrew).

Reznick, R. (2000b) 'Ministry demands medical records in human egg trading investigation', *Haaretz* (23 February), available at: http://www.haaretz.com (accessed 23 January 2001).

Reznick, R. and N. Kraw (2000) 'House arrest for two doctors suspected of fraud and endangerment of lives', *Haaretz* (16 May), available at: http://www.haaretz.co.il (accessed 23 January 2001) (Hebrew).

Reznick, R. and H. Shadmi (2001a) 'Professor Ben Raphael to be suspended from administration of the Women's Wing of Beilinson Hospital', *Haaretz* (22 January), available at: http://www.haaretz.co.il (accessed 23 January 2001) (Hebrew).

Reznick, R. and H. Shadmi (2001b) 'The Solicitor General will soon decide whether to rule against the doctors in the episode of the eggs', *Haaretz* (22 January) (Hebrew).

Roberts, E. F. S. (2007) 'Extra embryos: the ethics of cryopreservation in Ecuador and elsewhere,' *American Ethnologist*, 34, 1: 181–99.

Roberts, E. F. S. (2012) *God's Laboratory: Assisted Reproduction in the Andes* (Berkeley, CA: University of California Press).

Roberts, E. F. S. and N. Scheper-Hughes (2011) 'Introduction: medical migrations', *Body & Society*, 17, 2–3: 1–30.

Rozenblum, S. (2000a) 'The suspicion: doctors extracted eggs and sold them to other women', *Yediot Aharonot* (9 March) (Hebrew).

Rozenblum, S. (2000b) 'Women fear commerce in eggs: thousands of couples wait in vain for a donation', *Yediot Aharonot*, available at: http://www.ynet.co.il (accessed in 2002) (Hebrew).

Rozenblum, S. (2000c) 'The recommendation: to permit egg donation from deceased women', *Yediot Aharonot* (18 December) (Hebrew).

Rylko-Bauer, B., et al. (2006) 'Reclaiming applied anthropology: its past present and future', *American Anthropologist*, 108, 1: 178–90.

Scheper-Hughes, N. (2002) 'The ends of the body: commodity fetishism and the global traffic in organs', *SAIS Review of International Affairs*, 22, 1: 61–80.

Scheper-Hughes, N. (2009) 'Making anthropology public', *Anthropology Today*, 25, 4: 1–3.

Seruv (Courage to Refuse) (2012) Available at: http://www.seruv.org.il (accessed 18 September 2012).

Shafir, G. (1999) 'Zionism and colonialism, a comparative approach', in I. Pappé (ed.), *The Israel/Palestine Question* (London: Routledge), pp. 72–85.

Shenfield, F. (2011) 'Implementing a good practice guide for CBRC: perspectives from the ESHRE Cross-Border Reproductive Care Taskforce', *Reproductive Biomedicine Online*, 23, 5: 657–64.

Shmerling, G. (2000) 'A committee has been formed on the topic of fertilization and use of ova', *Ma'ariv*, 29 August (Hebrew).

Shohat, E. (1989) *Israeli Cinema: East/West and the Politics of Representation* (Austin, TX: University of Texas Press).

Shohat, E. (1997) 'Sephardim in Israel: Zionism from the standpoint of its Jewish victims', in A. McClintock, A. Mufti and E. Shohat (eds), *Dangerous Liaisons: Gender, Nation, and Postcolonial Perspectives* (Minneapolis, MN: University of Minnesota Press), 39–68.

Shteiber, O. (2002) 'All the eggs in one basket', *To Be a Family Magazine*, April: 13 (Hebrew).

Siegel, J. (2001) 'Health Minister approves Bill aimed at reducing shortage of donated ova', *Jerusalem Post*, available at: http://www.jpost.com/Editions/2001/03/20/News/News.23244.html (accessed 19 November 2001).

Siegel-Itzkovitch, J. (2000) 'Health Ministry investigating claims of illegal ova sales', *Jerusalem Post*, available at: http://www.jpost.com/Editions/2000/02/10/Health/Health.443.html (accessed 23 January 2001).

Snowden, R., G. D. Mitchell and E. M. Snowden (1983) *Artificial Reproduction: A Social Investigation* (London: George Allen and Unwin).

Sperling, D. (2010) 'Commanding the "Be fruitful and multiply" directive: reproductive ethics, law, and policy in Israel', *Cambridge Quarterly of Healthcare Ethics*, 19, 3: 363–71.

Spivak, G. (1987) *In Other Worlds: Essays in Cultural Politics* (New York: Methuen).

Stop the Wall, Palestinian Grassroots Anti-Apartheid Campaign (2011) Available at: http://www.stopthewall.org/ (accessed 16 August 2012).

Storrow, R. (2005) 'Quests for conception: fertility tourists, globalisation and feminist legal theory', *Hastings Law Journal*, 57, 2: 295–330.

Storrow, R. F. (2011) 'Assisted reproduction on treacherous terrain: the legal hazards of cross-border reproductive travel', *Reproductive Biomedicine Online*, 23, 5: 538–45.

Strathern, M. (1988) *The Gender of the Gift* (Los Angeles, CA: University of California Press).

Strathern, M. (1991) *Partial Connections* (Oxford: Rowman and Littlefield Publishers).

Strathern, M. (1992a) *After Nature: English Kinship in the Late Twentieth Century* (Cambridge: Cambridge University Press).

Strathern, M. (1992b) *Reproducing the Future: Essays on Anthropology, Kinship and the New Reproductive Technologies* (New York: Routledge).

Strathern, M. (1995) 'Displacing knowledge: technology and the consequences for kinship', in F. Ginsburg and R. Rapp (eds), *Conceiving the New World Order* (Berkeley, CA: University of California Press), pp. 346–68.

Teman, E. (2003) 'The medicalization of "nature" in the "artificial body": surrogate motherhood in Israel', *Medical Anthropology Quarterly*, 17, 1: 78–98.

Teman, E. (2010) *Birthing a Mother: The Surrogate Body and the Pregnant Self* (Berkeley, CA: University of California Press).

Thompson, C. (2005) *Making Parents: The Ontological Choreography of Reproductive Technologies* (Cambridge, MA: MIT Press).
Thompson, C. (2007) 'Why we should, in fact, pay for egg donation', *Regenerative Medicine*, 2, 2: 203–9.
Throsby, K. (2004) *When IVF Fails: Feminism, Infertility and the Negotiation of Normality* (London: Palgrave Macmillan).
Titmuss, R. (1970) *The Gift Relationship* (London: Allen Unwin).
Tremayne, S. (2006) 'Not all Muslims are luddites', *Anthropology Today*, 22, 3: 1–2.
Trif, A. (2005) 'Explaining diversity in industrial relations at company level in Eastern Europe: evidence from Romania', Dicussion Paper 05/0, Max Plank Institute for the Studies of Societies, Cologne, available at: http://www.mpifg.de/pu/mpifg_dp/dp05-3.pdf (accessed 13 August 2012).
Trif, A. (2008) 'Opportunities and challenges of EU accession: industrial relations in Romania', *European Journal of Industrial Relations*, 14, 4: 461–78.
Trouillot, M.-R. (2001) 'The anthropology of the state in the age of globalization: close encounters of the deceptive kind', *Current Anthropology*, 42: 125–38.
Verran, H. (1998) 'Reimagining land ownership in Australia', *Postcolonial Studies*, 1, 2: 237–54.
Vertovec, S. (2011) 'The cultural politics of nation and migration', *Annual Review of Anthropology*, 40: 241–56.
Visweswaran, K. (1994) *Fictions of Feminist Ethnography* (Minneapolis, MN: University of Minnesota Press).
Wade, P. (ed.) (2007) *Race, Ethnicity and Nation: Perspectives from Kinship and Genetics* (New York and Oxford: Berghahn Books).
Waldby, C. and M. Cooper (2010) 'From reproductive work to regenerative labour the female body and the stem cell industries', *Feminist Theory*, 11, 1: 3–22.
Wallace, C. (1999) 'Crossing borders: mobility of goods, capital and people in the Central European region', in A. Brah et al. (eds), *Global Futures: Migration, Environment and Globalization* (London: Palgrave), pp. 185–209.
Weiss, M. (1994) *Conditional Love: Parents' Attitudes toward Handicapped Children* (Westport, CT: Bergin & Garvey).
Weiss, M. (2002) *The Chosen Body: The Politics of the Body in Israeli Society* (Stanford, CA: Stanford University Press).
Weizman, E. (2007) *Hollow Land: Israel's Architecture of Occupation* (London and New York: Verso).
West, C. (1990) 'The new cultural politics of difference', in R. Ferguson (ed.), *Out There: Marginalization and Contemporary Cultures* (Cambridge, MA: MIT Press).
White, H. (1987) *The Content of the Form: Narrative Discourse and Historical Representation* (Baltimore, MD: Johns Hopkins University Press).
Widdows, H. (2009) 'Border disputes across bodies: exploitation in trafficking for prostitution and egg sale for stem cell research', *Journal of Feminist Approaches to Bioethics*, 2, 1: 5–24.
Wimmer, A. and N. Glick-Schiller (2002) 'Methodological nationalism and beyond: nation-state building, migration and the social sciences', *Global Networks*, 2, 4: 301–34.
Wolf, M. (1992) *A Thrice Told Tale: Feminism, Postmodernism and Responsible Ethnography* (Stanford, CA: Stanford University Press).

Yuval-Davis, N. (1989) 'National reproduction and the "demographic race" in Israel', in N. Yuval-Davis and F. Anthias (eds), *Woman–Nation–State* (London: Macmillan Press), pp. 92–109.

Yuval-Davis, N. (1997) *Gender and Nation* (London: Sage Publication).

Zimmerman, E., et al. (2010) 'Europe and Israel: an examination of a scientific path increasingly travelled together', in D. Preda (ed.), *The Road Europe Travelled Along: The Evolution of the EEC/EU Institutions* (Brussels: Peter Lang), 479–500.

Zuhur, S. (1992) 'Of milk-mothers and sacred bonds: Islam, patriarchy, and new reproductive technologies', *Creighton Law Review*, 25, 5: 1725–38.

Index

9780230319295